# Molecular Exploitation of Apoptosis Pathways in Prostate Cancer

# Molecular Medicine and Medicinal Chemistry

**Book Series Editors:**   Professor Colin Fishwick *(School of Chemistry, University of Leeds, UK)*
Dr Paul Ko Ferrigno and Professor Terence Rabbitts FRS, FMedSci *(Leeds Institute of Molecular Medicine, St. James's Hospital, UK)*

---

*Published:*

**MicroRNAs in Development and Cancer**
*edited by Frank J. Slack (Yale University, USA)*

**Merkel Cell Carcinoma: A Multidisciplinary Approach**
*edited by Vernon K. Sondak, Jane L. Messina, Jonathan S. Zager, and Ronald C. DeConti (H Lee Moffitt Cancer Center & Research Institute, USA)*

**DNA Deamination and the Immune System: AID in Health and Disease**
*edited by Sebastian Fugmann (National Institutes of Health, USA), Marilyn Diaz (National Institutes of Health, USA) and Nina Papavasiliou (Rockefeller University, USA)*

**Antibody Drug Discovery**
*edited by Clive R. Wood (Bayer Schering Pharma, Germany)*

**Molecular Exploitation of Apoptosis Pathways in Prostate Cancer**
*by Natasha Kyprianou (University of Kentucky, USA)*


*Forthcoming:*

**Fluorine in Pharmaceutical and Medicinal Chemistry: From Biophysical Aspects to Clinical Applications**
*edited by Véronique Gouverneur (University of Oxford, UK) and Klaus Müller (F Hoffmann-La Roche AG, Switzerland)*

Volume 5

Molecular Medicine and
Medicinal Chemistry

# Molecular Exploitation of Apoptosis Pathways in Prostate Cancer

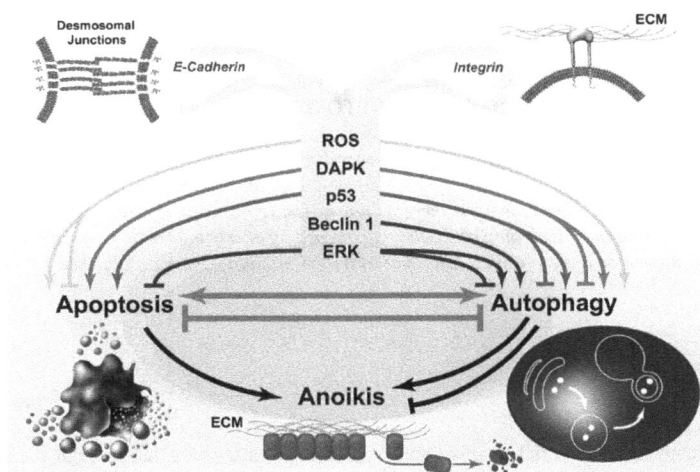

Natasha Kyprianou

University of Kentucky, USA

Imperial College Press

ICP

*Published by*

Imperial College Press
57 Shelton Street
Covent Garden
London WC2H 9HE

*Distributed by*

World Scientific Publishing Co. Pte. Ltd.
5 Toh Tuck Link, Singapore 596224
*USA office:* 27 Warren Street, Suite 401-402, Hackensack, NJ 07601
*UK office:* 57 Shelton Street, Covent Garden, London WC2H 9HE

**British Library Cataloguing-in-Publication Data**
A catalogue record for this book is available from the British Library.

**Molecular Medicine and Medicinal Chemistry — Vol. 5**
**MOLECULAR EXPLOITATION OF APOPTOSIS PATHWAYS
IN PROSTATE CANCER**

ISBN-13 978-1-84816-449-9
ISBN-10 1-84816-449-1

Printed in Singapore.

*For all the men fighting prostate cancer*

*and*

*in memory of those who lost the battle*

21  αἱ μὲν περιρρήξιες ταχεῖαι, αἱ δὲ τῶν ὀστέων ἀποπτώ-
    σιες, ᾗ ἂν τὰ ὅρια τῆς ψιλώσιος ᾖ, ταύτῃ ἀποπίπτουσι,378
    βραδύτερον δέ.* χρὴ δὲ τὰ κατωτ|έρω τοῦ τρώματας
    καὶ τοῦ σώματος τοῦ ὑγιέος προσαφαιρεῖν — προ-
    θνήσκει γὰρ — φυλασσόμενον ὀδύνην· ἅμα γὰρ λει- 5
    ποθυμίῃ θνήσκουσι. μηροῦ ὀστέον ἐκ τοιούτου ἀπε-
    λύθη ὀγδοηκοσταῖον, ἡ δὲ κνήμη ἀφῃρέθη εἰκοσταίη,

Copy of the original text in Chapter 35 of *ΜΟΧΛΙΚΟΝ* (378–376 BCE), a book
attributed to Hippocrates that deals with the instruments for the reduction
of fractures and dislocations, according to the transcription of Kuehlewein
(see Hippocrates, 1928). The word corresponding to apoptosis — in the plural
case — is boxed. In the translation below, this word is given in italics [*identifies
the end of the translation]. "In cases of fractured bones, lines of demarcation
form quickly, and the *falling off* of the bones occurs, where the limits of the
denudation may happen so that the bones are falling off, but more slowly."

# Preface

As we witness the end of the first decade of the 21st century, the questions we cancer researchers are most frequently asked are: when we are going to find the cure for cancer; and why we have not found it yet? At national and international conferences we see that data are interpreted at a more assertive rhythm and answers at the genetic and cellular level are generated at a similar pace, moving basic research forward into the oncology clinics and cancer centers globally. Recognized by the major funding agencies in the United States, Europe, Australia, and Japan as a top priority focus, cancer research drives translational applications from the bench to the clinic at a phenomenal rate, impacting therapeutic outcomes and affecting the lives of millions around the world. Yet, we still cannot declare with any degree of confidence to the scientific community that we have found the cure for cancer. The task is most challenging, the expectations are high, and the patience for clinically meaningful answers is running thin. These notions are translated by the public into a belief that powerful technology-driven research and long-term investment in the most talented scientific minds have yet to deliver cures, improve diagnosis, and increase patient survival. There is a general assumption that the complexity of cancer — both as a biological phenomenon and as a life-threatening disease — presents major obstacles in the fight to deliver an effective cure and maximally impact patient survival and quality of life after cancer; nevertheless there have been tremendous strides in research that lead to new treatments. An understanding that uncontrolled tumor growth is due not only to aberrant cell proliferation, but also to an imbalance between cell proliferation and programmed cell death, opened many exciting avenues for therapeutic investigations by selectively targeting apoptosis among the tumor cells.

Apoptosis, identified as a biological process vital to the well-being of an organism and defined as a genetically programmed phenomenon less than forty years ago, holds tremendous promise for the treatment of cancer in its molecular core. The characteristic analogy that gained popular use in the 1980s and early 1990s was the description of the "falling of the leaves from the trees in the autumn" (from the Greek for "falling from": *apo ptosis*). When I joined the John Isaacs laboratory at Johns Hopkins Oncology Center in 1986, the genetically regulated phenomenon of programmed cell death was being investigated at the mechanistic level in thymocytes after γ-irradiation or glucocorticoid treatment, by a pioneer Russian scientist, Dr Umansky (Umansky, Korol and Lelipovich, 1981; Umansky, 1982). This was distinctly different from the non-specific, randomly occurring process of necrosis that was not activated by any death stimulus but rather reflected the "collapse" of the cell in a non-programmed manner. The only measure of activation of programmed cell death was the fragmentation of DNA into nucleosomal-sized fragments. My research task was to determine whether androgen deprivation of the normal and malignant prostate leads to activation of programmed cell death and tissue involution. It was the beginning of a most exciting journey: the exposure to an intellectually rich environment, working with a brilliant mentor, to find ways to kill cancer cells and to promote the development of therapeutic regimes to treat patients with prostate cancer. More than two decades later, those inaugural experiences in activating cell-suicidal mechanisms for anti-tumor action in human cancers gave rise to molecular advances in the field of apoptosis, licensing investigators to explore beyond the boundaries of the current understanding of cancer development and clinical management of this life-threatening disease. Prostate cancer is a significant burden on men's health. It is a major cause of mortality in American men and a major contributor to all male cancer mortality in the United States and Western countries. Although patients with localized prostate cancer can survive for a long period, a large percentage of prostate cancer patients are diagnosed with advanced metastatic disease, or develop skeletal metastases. Prostate cancer metastasis to the skeleton and bone dramatically lowers the patient's routine activities and lifestyle and severely threatens his life. Castration-resistant prostate cancer (CRPC) (formerly known as androgen-independent prostate cancer) is the lethal stage of the disease and responsible for 30,000 deaths each year in the United States. This monograph was written in the hope and anticipation of translating current understanding of the molecular regulation of programmed mechanisms of cell death into a discovery platform, to allow pharmacologic exploitation of effective therapeutic modalities to conquer metastatic, castration-resistant prostate cancer.

# Contents

# 1

# Introduction: Prostate Cancer

Normal cellular physiology is a tightly regulated process, with positive and negative feedback loops that decide whether a cell should differentiate, divide, adapt to its microenvironment or commit suicidal death via apoptosis. The biochemical events controlling the cellular responses during normal organ homeostasis and disease development involve post-translational modifications of newly synthesized proteins; functional interactions between enzymatic and structurally scaffolding proteins; sequestration within specialized subcellular compartments; intracellular transport; and secretion into the extracellular space targeting key biological processes such as cell growth, apoptosis, neovascularization, and membrane trafficking. Alterations in cell metabolism affect development of human disease; this is the fundamental principle driving biomedical research and discovery.

Cancer development is a multistep process, involving multiple genetic alterations that drive normal cells into a highly malignant, multicompetent phenotype. Hanahan and Weinberg (2000) eloquently proposed that tumorigenesis involves six essential physiological processes becoming deregulated in terms of cell behavior: self-sufficiency in growth signals, insensitivity to growth-inhibitory (anti-growth) signals, evasion of apoptosis, limitless replicative potential, sustained angiogenesis, and tissue invasion and metastasis. In each of these steps, the tumor microenvironment is a critical participant, through the extracellular matrix (ECM) maintenance and remodeling as well as cell–cell interactions. The striking ability of a tumor cell population to grow exponentially represents an imbalance between cellular proliferation and cell death, in favor of cell proliferation and at the expense of cell death, but growth pressure alone will not cause a mass of cells to be malignant. While enjoying extensive communications with the ECM, cancer cells migrate autonomously across tissue boundaries, and have the audacious capacity to survive and grow among foreign cell populations. The true life-threatening behavior of malignant cells is their propensity to infiltrate and usurp the "sovereignty" of host tissue societies. Although some Mendelian cancer susceptibility genes are involved in tumor initiation, others probably act at later stages in tumorigenesis. Most Mendelian susceptibility genes are tumor suppressors that cause dominantly inherited disease. The protein products of these genes are inactivated by a "second hit", often a loss of heterozygosity in somatic progenitor cells. Another classification divides the cancer susceptibility genes into gatekeepers (genes that regulate essential cellular functions such as growth and apoptosis control), caretakers (genes that maintain genomic integrity), and landscapers (genes that restrain tumorigenesis by providing the suitable microenvironment) (Kinzler and Vogelstein, 1998).

An overwhelming body of evidence suggests that the ability of tumor cells to evade programmed cell death — or apoptosis — is possibly the major molecular force driving the initiation, growth, and progression of human tumors to metastatic disease (reviewed by Mehlen and Puisieux, 2006). Apoptotic evasion indeed represents one of the true hallmarks of cancer and appears to be a vital component in the immunogenic, chemotherapeutic, and radiotherapeutic resistance that characterizes the most aggressive of human cancers including prostate cancer (Westin, Bergh and Damber, 1993; McKenzie and Kyprianou, 2006). It is no surprise therefore that molecular and genetic manipulation of apoptotic signaling pathways dominates clinical investigations globally in the treatment of human cancer, the prominent killer disease associated with impaired apoptosis (Bjornsti and Houghton, 2004a). A rapidly expanding series of biochemical and genetic studies have substantially increased our understanding of death signal transduction pathways, making it clear, however, that apoptosis is not a single-lane, one-way street. Rather, multiple parallel pathways have been identified, primarily led by the caspase family of proteases towards the controlled demolition of cells (Taylor, Cullen and Martin, 2008). For instance, while establishing distinct roles for each of these apoptotic players, analysis of BCL-2, BAX, p53, and caspase knockout mice also provided valuable information for the design of specific inhibitors of apoptosis (Veis *et al.*, 1993; Kamada *et al.*, 1995; Kuida *et al.*, 1998). Functional studies indicate with rigor and conviction that by blocking one executioner pathway, as in caspase knockout mice, what we observe is not a complete suppression of apoptosis but rather a delay in apoptosis induction (Yoshida *et al.*, 1998). Recently, an elegant model of proteolytic reciprocal negative regulation has been proposed that opens a promising mechanistic basis for the combined therapeutic use of proteosome inhibitors and pro-apoptotic agents targeting specific caspases (Gray, Mahrus and Wells, 2010). In view of nature's means of ensuring activation of a compensatory apoptotic response, when one pathway fails in developing prostate cancer therapeutic interventions, the challenge remains to dissect individual apoptotic pathways further. This evidence will be reviewed and discussed here in the context of its translational

significance in the development of therapeutic intervention approaches strategi-
cally targeting prostate cancer.

Cancer is lethal in large part because of its ability to metastasize to remote
locations. Indeed, if metastasis did not exist as a phenomenon, and every solid
neoplasm was simply a matter of resection and management of local recurrences, the term "cancer" would possibly lose much of its strong negative impact
on health. Despite the medical community's most assiduous attempts at early
detection before metastatic capability is acquired, many cancers (such as ovarian
carcinoma) either cannot be detected effectively at an early stage, or frequently
metastasize at an early stage of development (for example, melanoma). Even in
cancers amenable to early detection, such as localized prostate cancer, many
patients still develop metastatic disease, ultimately proving fatal. On progression to high-grade neoplasia, transformed cells acquire qualities that enable
them to invade neighboring tissues and ultimately to metastasize to distant
sites. The cellular events involved in tumor cell progression to metastasis
include loss of cell adhesion, increased cell motility and invasion, entry into the
circulation and survival during the journey, extravasation, and colonization of
distant sites (Gupta and Massague, 2006). These events are largely mediated by
cadherins, which function to maintain adherens junctions between cells, and
integrins, which upon engagement with components of the extracellular matrix,
reorganize to form adhesion complexes (Fornaro, Manes and Languino, 2001).
Loss of expression of cadherins ostensibly promotes metastasis by enabling the
first step of the metastatic cascade: loss of functional intercellular adhesion
junctions and consequent disaggregation of cancer cells from one another is
a requirement for tumor cells to invade locally and to metastasize to distant
organs (Takeichi, 1991). Integrins are expressed by every cell type derived
from the three primary germ layers, and their presence is required for the
maintenance of normal tissue architecture by functioning as receptors for
ECM proteins, including collagen, laminin, and fibronectin (Hynes and Lander,
1992). Several other structural proteins associated with adherens junctions can
also mediate intracellular signaling functions in human cancers (Hiroashi,
1998; Perez-Moreno and Fuchs, 2006). At the biochemical level, the mechanism
for initial invasion may parallel that used by non-malignant cells that traverse
tissue boundaries. Investigative efforts using specialized cell models and
powerful array technology are quickly uncovering the functional interplay of
signaling effector molecules that mediate the malignant state of cancer cells
(Gingrich *et al.*, 1997; Clark *et al.*, 1999; Zhu *et al.*, 2006).

The immense therapeutic significance associated with the molecular
targeting of apoptotic players during cancer progression to metastasis is clearly
recognized as a top priority in drug-discovery efforts. Resistance to apoptosis
is not only critical in conferring therapeutic failure to standard treatment
strategies, but anoikis (apoptosis upon loss of anchorage and detachment from

the ECM) also plays a critically important role in angiogenesis and metastasis. Indeed, it is the very ability to survive in the absence of adhesion to the ECM and after losing contact with neighboring cells (following disruption of adherens junctions), that enables tumor cells to disseminate from the primary tumor site, invade a distant site, and establish a metastatic lesion. Tumor cells can escape from detachment-induced apoptosis by tightly controlling anoikis pathways, including the extrinsic death receptor pathway and the ECM-integrin mediated cell survival pathways (Frisch *et al.*, 1996; Frisch and Screaton, 2001; Bix and Iozzo, 2005; Renebeck, Martelli and Kyprianou, 2005). Considering the functional promiscuity of individual signaling effectors, it is critical to dissect the mechanistic networks driving tumor cells to evade anoikis and embark on a metastatic spread (Sakamoto and Kyprianou, 2010). Resistance to death via anoikis dictates tumor cell survival and provides a molecular basis for therapeutic targeting of metastatic disease.

Prostate cancer is becoming an epidemic in North America and Western nations. In the United States, 86 men die from prostate cancer each day. A malignancy that claims the lives of thousands of patients remains the second leading cause of cancer-related death in the United States and the most frequently diagnosed non-cutaneous malignancy in men. It is estimated that in 2010, 192,000 new cases will be diagnosed and that more than 27,000 patients will succumb to the disease. The statistics are alarming but the clinical scenario is optimistic: although the clinical course of newly diagnosed prostate cancer cannot be predicted with certainty, most will turn out to be relatively slow-growing, small, low-grade and non-invasive, and posing little risk to the life or health of the host. For prostate cancer patients with clinically localized disease, treatment by surgery or radiation typically can be curative. Seventy years ago, Charles Huggins first advocated androgen ablation monotherapy through medical or surgical castration for the treatment of prostate cancer patients (Huggins and Hodges, 1941). This was the first recognition of the hormonal dependence of prostate cancer, setting the endocrine landscape for prostate tumors that persists to the present (Jin *et al.*, 1996; Hsing, 2001). In the year 2010, androgen ablation remains a primary strategy to combat metastatic disease, with the underlying mechanism driving the therapeutic response being apoptosis induction among the androgen-dependent prostate cancer epithelial cells (Kyprianou, English and Isaacs, 1990; Colombel *et al.*, 1993; Westin *et al.*, 1995).

While investigating the impact of steroid biosynthesis by gonads and adrenal glands on prostate growth and the cellular distribution of the key metabolites in the prostate gland, John Isaacs made the groundbreaking observation that androgens play a critical role in the growth of androgen-dependent prostate tumors by blocking cell death (Isaacs, 1987). Subsequent work in the Isaacs laboratory in the late 1980s identified apoptosis as the predominant form of tumor cell demise caused by both androgen ablation and chemotherapeutic agents (Kyprianou and Isaacs, 1988a, 1989; Kyprianou, English and Isaacs, 1990), intracellularly triggered by transforming growth factor-β (TGF-β) (Kyprianou and Isaacs, 1988b). Resembling a "portal", in time these observations identified prostate apoptosis induction by androgen deprivation as the molecular mechanism driving the therapeutic response of prostate cancer patients (Kyprianou, English and Isaacs, 1990). This was documented by the pioneering work of Dr Huggins, showing that removing the androgen source via castration will regress the prostate tumor and cure the patient, at least initially (Huggins, Stevens and Hodges, 1941). A few years later, at the time when the mechanism of apoptosis was being investigated eagerly by several groups, we reported a series of studies indicating that apoptosis is of major significance in driving the therapeutic response of prostate patients to radiotherapy, and plays a critical role in prostate tumor radiosensitivity (Sklar *et al.*, 1993; Kyprianou *et al.*, 1997; Szostak and Kyprianou, 2000), via caspase activation (Winter, Rhee and Kyprianou, 2004). Androgen deprivation therapy still remains the only therapeutic option that can positively influence the course of the disease, once it is too advanced for local treatment. Although reducing circulating androgen levels is initially effective in eliciting cell death and in reducing the cancer burden, this strategy affords a median patient survival of less than 36 months (Glass *et al.*, 2003). Prostate cancer cell survival may be maintained due to the fact that current hormone ablation treatment regimens, although effective in reducing circulating levels of hormone, only decrease tissue androgen levels by 75% (Mostaghel, Page and Lin, 2007). It is the acquisition of anti-apoptotic signal transduction that ultimately leads to the characteristic treatment resistance that typifies advanced prostate cancer. Considerable efforts have been expended since then in the apoptosis arena, to gain a better understanding of the molecular targets and mechanisms contributing to the emergence of castration-resistant prostate tumors, and a litany of studies investigating mechanisms of therapeutic failure has paraded the literature (Murphy, Soloway and Barrows, 1991; Blackledge, 2003; Bogdanos *et al.*, 2003). Further dissection, at the molecular level, of the apoptotic signaling pathways utilized by prostate tumors during disease progression will lead to novel therapeutic targets. These will combat metastatic prostatic cancer and be independent of, or work in combination with, current hormonal ablation regimens. The challenge remains to effectively eliminate androgen-dependent and androgen-independent prostate tumors during any stage of progression by recognizing that the timing of androgen deprivation therapy and chemotherapy in patients with advanced prostate cancer is fundamentally important in establishing high therapeutic response (Isaacs, 1984). Indeed, overcoming the androgen-independence of prostate tumors is considered the most critical therapeutic endpoint for improving patient survival. Radical prostatectomy, androgen ablation monotherapy and radiation therapy are considered curative for localized prostate cancer, but no effective treatment for patients with metastatic disease is currently available (Szotak and Kyprianou, 2000; Denmeade and Isaacs, 2002). There are two primary contributors in the emergence of androgen-independent metastatic prostate cancer: activation of survival pathways, including apoptosis suppression and anoikis resistance; and increased tumor neovascularization (Hannahan and Weinberg, 2000; McKenzie and Kyprianou, 2006). Consequently, targeting of apoptotic players is of vital therapeutic significance since resistance to apoptosis is not only critical in conferring therapeutic failure to standard treatment strategies, but anoikis resistance also plays a paramount role in tumor progression to metastasis (Frisch and Screaton, 2001; Sakamoto and Kyprianou, 2010).

Active surveillance may be appropriate for many men with a favorable (low risk) cancer, so the means to distinguish indolent cancers from those that are potentially lethal, before initiating treatment, would be a significant benefit to these men. However, for the many patients in whom there is recurrence or diagnosis with metastatic disease, the primary goal becomes not only to provide treatment options to effectively cure the tumor, but to also provide long-term disease control with minimal compromise of the quality of life. The latter entails hormone-based therapeutic strategies that suppress androgens and the androgen receptor (AR) signaling axis (Bonaccorsi *et al.*, 2004a), as well as the directed activation of survival mechanisms, by targeting epidermal growth factor (EGF) receptor interactions (Bonaccorsi *et al.*, 2004b; Farhana *et al.*, 2004). Surgical resection of locally invasive prostate cancer yields a 10-year progression-free survival probability of greater than 10%, while the development of metastatic prostate cancer dramatically reduces the survival probability to 35% (Glass *et al.*, 2003). Unfortunately these recurrent, "castration-resistant" cancers (CRPC) represent the ultimately lethal phenotype of the disease (Feldman and Feldman, 2001). The striking requirement for AR activity promoted the efforts for its suppression via androgen ablation and/or use of direct AR antagonists towards therapeutic efficacy monitored by tumor regression and palliation of symptoms. The durability of the clinical effect ranges from months to years, and tumor regrowth and metastatic progression occurs.

The dynamics of prostate cancer characterization and detection rate took a dramatic turn since the introduction of prostate-specific antigen (PSA) in 1986. Prostate cancer screening, using serum testing for PSA along with digital rectal examination has become commonplace in the developed world, with most men in the United States over the age of 50 having been screened. PSA, a protein secreted by prostatic epithelial cells to become a component of the ejaculate, appears in the circulation only as a result of epithelial damage or prostate dysfunction. As such, although PSA testing is frequently used as a screening tool, in this setting, serum PSA levels are not biomarkers of prostate cancer *per se*. Rather, the serum PSA can be elevated in a variety of non-cancerous prostate diseases and conditions, including prostate infections, symptomatic and asymptomatic non-bacterial prostatitis, and benign prostatic hyperplasia/hypertrophy (BPH). Considering that many prostate tumors currently detected may be of an indolent nature, a current challenge is identifying the patient who may benefit from treatment and/or in whom side effects of treatment may be acceptable. Because inflammatory conditions and/or BPH affect more than half of the men in the prostate cancer screening age group, the significance of an isolated elevation in serum PSA is often neither clear nor easily interpreted with any level of confidence.

At the histopathological level, prostate cancer is a heterogeneous tumor, with a natural history of progression from prostatic intraepithelial neoplasia (PIN) to locally invasive androgen-dependent to androgen-independent metastatic disease (Arnold and Isaacs, 2002; Debes and Tindal, 2004). Compelling evidence supports that this tumor progression to androgen-independent metastatic disease results from roadblocks in the apoptotic pathways, rather than from uncontrolled cell proliferation (Djakiew *et al.*, 1998; DiPaola, Patel and Rafi, 2001; Goswami *et al.*, 2005; McKenzie and Kyprianou, 2006); moreover metastatic prostate cancer cells have a remarkably low rate of cell proliferation (<5% of cells proliferating in metastatic sites). The low proliferative rate is mainly responsible for the relative unresponsiveness of human prostate tumors to standard anti-proliferative chemotherapy, while highly proliferative androgen-independent prostate cancer cells remain exquisitely sensitive to apoptosis, despite their acquisition of unique dependence on different signals for survival. To reinstate these apoptotic pathways and impair the active survival signaling pathways, there has been tremendous investment in efforts to optimize therapeutic agents acting through the intrinsic mitochondrial, extrinsic death receptor pathways or endoplasmic reticulum (ER) stress pathways to induce apoptosis in prostate cancer cells (Schwarze *et al.*, 2008). However, androgens, and the active metabolite dihydrotestosterone (DHT), do not have monopoly of the prostate growth control as was originally believed (Gloyna, Siiteri and Wilson, 1970). The explosive growth of evidence in the last two decades firmly established that the landscape of growth tumorigenesis has a multilayered complexity due to endocrine and paracrine contributions. Androgens primarily control prostate growth via well-orchestrated kinetics to maintain homeostasis (Isaacs, 1987; Kyprianou and Isaacs, 1988a). The functional involvement of several cytokines activates diverse molecular mechanisms that sustain the growth equilibrium of the prostate gland. An imbalance in the growth kinetics in the aging gland results from activation of signaling pathways that promote proliferation (EGF, HER2/neu) (Pollak, Beamer and Zhang, 1998; Djakiew, 2000; Ornitz and Itoh, 2001; Torring *et al.*, 2003), and loss of signaling events that activate apoptosis (androgens, AR) (Isaacs, 1987; Saric and Shain, 1998; Gregory *et al.*, 2004; Knudsen and Scher, 2009) during progression to metastasis (Rosini *et al.*, 2002). TGF-β plays a most prominent role in the control of prostate growth dynamics independent of androgens. This player is multi-faceted (functionally), multilayered (cell type distribution), and multi-regulated (interaction with other signaling effectors). From its role as an apoptosis inducer (Kyprianou and Isaacs, 1989; Guo and Kyprianou, 1999) and anoikis promoter (Ramachandra *et al.*, 2002); a mediator of a rigorous cross-talk with the androgen signaling axis (Bruckheimer and Kyprianou, 2001; Hayes *et al.*, 2001; Song *et al.*, 2003) and epithelial–mesenchymal transition (EMT) (Thiery, 2002; Zavadil and Bottinger, 2005; Zhu and Kyprianou, 2010), and invasion (Hjelmeland *et al.*, 2005) recognized during tumor progression to metastasis, it can be seen that this cytokine is a fully-charged "bullet" of prostate tumor growth governance. The progression to metastasis is a complex process and involves numerous biological functions that collectively enable cancer cells from a primary site (the prostate gland) to disseminate and overtake distant organs (lymph nodes, liver). However, prostate tumors engage in a seductive genetic alteration dynamic and it would be naïve to believe that restoring AR activity and suppressing tumor growth under conditions of androgen deprivation would lead to successful combinatorial therapies. Antibody targeting of interleukin-6 (IL-6) results in a significant suppression of prostate tumor growth *in vivo* (Smith and Keller, 2001). Molecular targeting of apoptotic signaling pathways has been heavily exploited in recent years by big pharmaceutical companies and academic investigators, with efforts directed towards the development of effective therapeutic modalities for the treatment of patients with advanced androgen-independent prostate tumors.

Most normal epithelial and endothelial cells undergo anoikis when they become detached from the ECM (Frisch and Ruoslahti, 1997). In contrast, prostate tumor cells are capable of evading anoikis and metastasizing to distant organs and selectively within the bone (Rennebeck, Martelli and Kyprianou, 2005; Hall *et al.*, 2006; Sakamoto and Kyprianou, 2010). Is the biological repertoire of the epithelial and endothelial cells sufficient to account for the events associated with the process of anoikis during prostate cancer metastasis? While there is no clear answer to this question, it is increasingly evident from existing knowledge that molecules that induce anoikis in tumor epithelial and endothelial cells provide exciting new leads into effective therapeutic targeting, as well as being markers of prostate cancer progression and prediction of thera-peutic efficacy/resistance. Prostate cancer cells have the ability to acquire both intracellular survival pathways, and alterations in chemokine and growth factor signal transduction, that allow them to circumvent either mode of cell death (apoptosis versus anoikis) or either signaling pathway of apoptosis (extrinsic versus intrinsic), and ultimately contribute to the androgen-independent and aggressive phenotype that is classically resistant to any form of conventional chemotherapy. Now, more than ever, it is critical to identify a predictive marker or molecular signature before or during pre-clinical anti-tumor drug development, to allow the selection of appropriate patients who are likely to respond, and exclusion of patients who are unlikely to exhibit any clinical benefit.

Overcoming the androgen-independence of prostate tumors is considered the most clinically critical therapeutic endpoint for improving the survival in patients with advanced metastatic prostate cancer. As one dissects the mechanisms underlying prostate cancer progression to hormone independence (CRPC) and treatment resistance during the clinical course of the disease (Gregory *et al.*, 2001a), the role of apoptotic evasion takes center stage. The ability of prostate cancer cells to activate intracellular survival pathways, coupled with the critically dynamic intracellular cross-talk and anti-apoptotic pathway redundancy, creates a formidable opponent for the most powerful of cytotoxic therapies, and even complete hormone withdrawal. Furthermore, the ability of these cells to adapt to their extracellular microenvironment by alterations in:

1. epithelial–stromal interactions;
2. pathophysiologic cellular stress responses;
3. growth factor-receptor pathways; or
4. the inflammatory response

allows the most hostile of tumor microenvironments to promote rather than inhibit cancer cell survival and, ultimately, encourage aggressive phenotypes. Whether it is the up-regulation of intracellular survival pathways, or the extracellular influence that up-regulates intracellular anti-apoptotic signal transduction that allows for such aggressive adaptation, remains a subject of debate. It is becoming increasingly apparent, however, that our ability to positively improve the therapeutic response and survival of patients with castration-resistant metastatic prostate cancer will ultimately require a therapeutic arsenal that targets multiple and often functionally overlapping signal transduction pathways rather than the current, frequently ineffective attempts at monotherapy for advanced disease. In light of the rapidly accumulating information which advances our understanding of the integrated functions governing cell proliferation and cell death, prostate cancer therapies which are not only molecularly targeted but that are also customized to take into account the delicate balance of opposing growth influences may be realized as the new millennium unfolds.

This monograph integrates the currently understood concepts governing the mechanisms of programmed cell death, and their therapeutic exploitation towards broadening of our existing apoptosis-based pharmacologic development to impair prostate cancer growth and progression to metastasis. The cellular demise can be induced by the "classic" apoptosis signaling to suppress tumor initiation and via anoikis, to block tumor cell migration, invasion and metastatic spread. Further detailed dissection of the apoptotic signaling pathways utilized by prostate tumors is expected to lead to the identification of new therapeutic targets to combat metastatic prostatic cancer that are independent of, or work in combination with, the current hormonal ablation regimens.

Most prostate tumors are initially responsive to androgen ablation therapy, which acts to deprive these tumor cells of their primary growth stimulus (Kyprianou, English and Isaacs, 1990; Colombel *et al.*, 1996). However, under the physiologic stress of hormone ablation, prostate tumors ultimately progress to androgen-independent disease that is resistant to both hormone ablation, as well as other systemic chemotherapies (McKenzie and Kyprianou, 2006). The challenges in the development of effective treatment modalities for advanced prostate cancer represent a classic paradigm of the functional significance of anti-apoptotic pathways in the development of therapeutic resistance.

Prostate cancer may metastasize and grow well in the bone because of the unique bone microenvironment in which the cancer cells find a "nurturing" niche to advance their aggressive phenotype and spread malignant growth. Prostate cancer bone metastases are comprised of both osteolytic and osteoblastic lesions. Under the intense stimulation of a plethora of bone matrix-derived growth factors, the balance of bone-forming and bone-destruction activities found in close proximity to the bone microenvironment favors maintaining the metastatic bone masses, the major cause of mortality in prostate cancer patients (Whitmore, 1984).

In the United States, African-American men suffer more than twice the incidence and twice the mortality from prostate cancer, compared to Caucasian men of European origin. Despite considerable etiologic research to identify genetic factors for prostate cancer and the racial disparity in the incidence of the disease (Thomas *et al.*, 2008), very limited new information has been gathered to provide clinically useful knowledge in racial differences that may impact prostate cancer development and progression (Carter *et al.*, 1992). In the 21st century there is unfortunately still a racial/ethnic disparity with regards to the treatment of clinically localized prostate cancer. As African-American and Hispanic men present with higher-grade (poorly differentiated) prostate cancer, they are less likely to receive definitive treatment. More importantly, this disparity in treatment exists among men diagnosed with poorly differentiated (high mortality risk) prostate cancer. If, in fact, definitive prostate cancer treatment improves prostate cancer survival, the uncovering of this treatment disparity may contribute to prostate cancer disparity and mortality. Epidemiologic evidence suggests that prostate cancer morbidity and mortality ought to be preventable among the different ethnic and racial groups. New insights into the molecular pathogenesis of prostate cancer offer new opportunities for the discovery of chemoprevention drugs and new challenges for their development. Established pathways that lead to the approval by the US Food and Drug Administration (FDA) of drugs for advanced prostate cancer may not be appropriate for the development of drugs for prostate cancer chemoprevention.

For example, large randomized clinical trials designed to test the efficacy of new chemoprevention drugs on prostate cancer survival in the general population, are likely to be conducted at great expense and take many years, threatening to increase commercial development risks while decreasing exclusive marketing revenues. As a consequence, to accelerate research progress, new validated surrogate and strategic clinical trial endpoints, and new clinical trial designs featuring more precisely defined high-risk clinical trial cohorts, are needed.

# 2

# The Prostate Gland Dynamics

The human prostate is a fibromuscular walnut-sized gland situated in the pelvis, and surrounds the proximal portion of the male urethra as it originates from the base of the bladder. The anatomy and development of the human prostate gland was first described at the beginning of the 20th century when Lowsley (1918) defined five lobes. Lowsley's work, based on fetal glands, gave the initial description of a dorsal or posterior lobe, a median and two lateral lobes, and additionally a ventral lobe which atrophied at birth. In the adult human prostate these lobes are fused and cannot be separated or defined by dissection, promoting controversy and long debates regarding the anatomic divisions characteristic of the prostate gland. It was not until the late 1970s that McNeal classified the prostate gland into a peripheral, transitional, central, and a periurethral zone, and this is the nomenclature most commonly used to describe the human prostate (McNeal, 1978). These zones contain histologically distinct ductal systems. The peripheral zone is the largest region of the glandular prostate and the principal anatomical origin for adenocarcinoma of the prostate (Cunha et al., 1983; McNeal et al., 1988). The transition zone represents the branches of the two most ductal distal lateral line duct pairs that empty into the urethra at the base of the verumontanum. Benign prostatic hyperplasia is commonly found in the transitional zone and in the periurethral region, leading to early obstructive urinary symptoms, and hence early detection (Che and Grignon, 2002). The central zone ducts follow the course of the ejaculatory ducts and empty into the urethra proximal to the verumontanum. The central zone is virtually, but not completely, disease-free.

Prostatic organogenesis is strikingly dependent upon reciprocal mesenchymal–epithelial interactions. Histologically, the prostate is composed of compound tubuloalveolar glands formed by a columnar or pseudostratified epithelium, interspersed in a rich fibromuscular stroma. The epithelium of the prostate develops from the ectoderm-derived epithelium of the lower portion of the urogenital sinus. Mesenchymal–epithelial interactions are exploited to determine the respective roles of epithelium and mesenchyme in the organogenic process, or to determine the respective roles of the epithelium and mesenchyme in the hormonal response of the gland (Cunha and Doncajour, 1989). The classic work by Cunha and colleagues (Cunha *et al.*, 1983; Cuhna *et al.*, 1996) defined that the outcome of epithelial–mesenchymal interactions is dependent not only upon the source of the inducing mesenchyme, but also on the germ layer origin, and on the responsiveness and age of the epithelium. The normal prostate consists of large glands with papillary infoldings that are lines with a two-cell layer consisting of basal cells and columnar secretory epithelial cells with pale cytoplasm and uniform nuclei (English, Santen and Isaacs, 1987; Evans and Chandler, 1987). The prostatic secretions are a complex mixture of biologically heterogeneous compounds secreted during ejaculation and believed to be of vital importance in the process of insemination. These prostate secretions include, but are not limited to, zinc, citrate, spermine, spermidine, phosphorycholine, cholesterol, lipids, prostate specific antigen, and prostatic acid phosphatase.

While mesenchymal–epithelial interactions play a fundamental role in the development of the prostate, the overall process of glandular development is elicited by androgens, which regulate prostatic development, growth, and function primarily via the androgen receptor (AR). During embryonic development, androgens induce the appearance of prostatic epithelial buds, ductal elongation, and branching. As prostatic differentiation proceeds, beginning at the prostatic urethra and extending distally towards the ductal tips, the periepithelial mesenchymal cells differentiate into AR-positive smooth muscle cells. Initially, the mesenchymal cells predominantly express vimentin, and as development progresses the mesenchymal cell population becomes highly organized into smooth muscle sheaths expressing desmin, myosin, and laminin (Cunha *et al.*, 1996).

The post-pubertal prostate is in a dynamic steady state, with cell death and cell proliferation being equal at approximately 1–2% per day (Berges *et al.*, 1995; Kyprianou, Tu and Jacobs, 1996). This delicate balance is maintained by normal physiologic levels of androgens in the adult prostate (Berry *et al.*, 1984; Hsing, 2001). Testosterone metabolism during embryonic development and in young adult life is a major activity driving the cellular responses in the prostate gland (Brasel, Coffey and Williams-Ashman, 1968; Siiteri and Wilson, 1974). Supraphysiologic levels of androgens in dogs and in habitual anabolic steroid users result in proliferation of the glandular epithelial cells in the prostate (Berry and Isaacs, 1984). The dramatic apoptosis of the prostate glandular epithelial cells in response to castration-induced androgen withdrawal is a vivid reminder of the striking dependence of the normal prostatic glandular epithelial cells on physiological androgen levels for their survival (English, Drago and

Santen, 1985; Isaacs, 1987; Kyprianou and Isaacs, 1988). The androgen-mediated survival of epithelial cells is mediated by paracrine factors secreted by the surrounding AR-positive stromal cells (Kurita *et al.*, 2001). The basal epithelial cells remain unaffected, contributing little to the involution of the prostate (English, Kyprianou and Isaacs, 1989; Buttyan *et al.*, 1999).

Prostate growth dynamics during aging involve disruption of the molecular mechanisms regulating apoptosis and cell proliferation among the stroma and epithelial cell populations, towards aberrant growth of the gland (Isaacs *et al.*, 1992; Kyprianou, Tu and Jacobs, 1996; Zeng *et al.*, 2004). In aging males, the two primary pathological conditions intimately linked to growth imbalance at the expense of apoptosis and in favor of cell proliferation, are benign prostate hyperplasia (BPH) and prostate cancer (Berry *et al.*, 1984; Kyprianou, Tu and Jacobs, 1996). This dramatic shift in cell growth and uncontrolled proliferative capacity characterizes the glandular epithelial as well as the basal cells of the human prostate gland in BPH patients (Kyprianou, Tu and Jacobs, 1996). Prostate cancer is a heterogeneous disease that progresses from prostatic intraepithelial neoplasia (PIN), to locally invasive, to castration-resistant metastatic carcinoma. Prostatic adenocarcinoma is recognized histopathologically through a combination of architectural and cytological features that provide the basis for the Gleason pattern score (Gleason, 1966). The pathological method of Gleason grading is of immense prognostic value to the urologist, as it correlates directly with several histopathologic and clinical endpoints, such as tumor size, margin status, clinical stage, disease progression, and patient survival (Gleason and Mellinger, 1974; Humphrey, 2004). Typically the malignant glands are smaller than the associated benign acini and are often arranged in infiltrative patterns that contrast with the more organized, lobular arrangement of the benign glands (McNeal, 1992). PIN is recognized as a precursor lesion for adenocarcinoma of the prostate and is represented by a spectrum of atypical epithelial changes that develops in pre-existing ducts and acini (Bostwick and Brawer, 1987). As in prostatic adenocarcinoma, nucleolar prominence is a key finding in high-grade PIN. In contrast to the acini of prostatic adenocarcinoma, the glandular structures of high-grade PIN have a residual basal cell layer, a feature that is fundamentally important in the differential diagnosis of prostate adenocarcinoma versus high-grade PIN (Bostwick and Brawer, 1987). At the same time solid morphological evidence links high-grade PIN with adenocarcinoma that includes ploidy values, genetic and cytogenetic changes, cytomorphology, morphometric values, lectin-binding patterns, and the topographical distribution in the prostate. Prostatic tumors are typically peripheral and often have a yellow to grey solid character that contrasts with the frequently spongy appearance of the adjacent non-malignant prostatic parenchyma.

Recent studies using protein microarray technology demonstrated that activation of PI3 kinase survival pathways and apoptosis suppression are early events in the microenvironment of prostate cancer growth (Paweletz *et al.*, 2001; Gao *et al.*, 2003; Espina *et al.*, 2006). Direct quantitative evidence has been generated to indicate that suppression of apoptosis in human PIN and invasive prostate cancer is associated with AKT phosphorylation and its substrate glycogen synthase kinase 3β (GSK3β) (Lin *et al.*, 2001; Liao *et al.*, 2003; Fizazi, 2007). This profound activation of AKT plays a causal role in prostate cancer progression to castration-resistant androgen-independent disease (Ghosh *et al.*, 2003; Stern, 2004). Mechanistically downstream components of the apoptotic cascade (cleaved and non-cleaved poly ADP-ribose polymerase, PARP) are shifted towards pro-survival messages at the cancer invasion front. High-grade PIN exhibited a lower level of phospho-extracellular signal-regulated kinase (ERK) compared with normal-appearing epithelium, while invasive prostate tumor epithelial cells contained even lower levels of phospho-ERK. Augmentation of the ratio of phosphorylated AKT to total AKT will suppress downstream apoptosis pathways through intermediate intracellular substrates such as GSK3β (Paweletz *et al.*, 2001). Differential evidence suggests that AKT suppresses androgen-induced apoptosis by directly phosphorylating and inhibiting the AR (Lin *et al.*, 2001). This reduction in apoptosis will directly shift the balance in favor of cell proliferation (and at the expense of cell death) among the glandular prostate epithelial cells, ultimately promoting the accumulation of cells within the prostate epithelium in massive numbers. Indeed, the high cell density and rapid and uncontrolled growth and size of cancer consequential to this severe disturbance of the prostate epithelium landscape are conventionally attributed to a pathologically high ratio of cell proliferation to cell death (Berges *et al.*, 1995; Bruckheimer and Kyprianou, 2000). Yet these features, although intimately linked to growth dynamics, might also result from inappropriate cell movement and inappropriate interactions with other cell types in the microenvironment (Chung *et al.*, 1989; Byrne, Leung and Neal, 1996; Fizazi *et al.*, 2003), also implicated in invasion and metastasis (Levine *et al.*, 1998; Rennebeck, Martelli and Kyprianou, 2005).

Malignant prostate cells escaping the primary tumor mass respond to host signals that call up the capacity for motility and survival (Frisch and Francis, 1994; Sakamoto and Kyprianou, 2010). Tumor epithelial cells can penetrate host cellular and extracellular barriers with the assistance of degradative enzymes produced by the host cells, but locally activated by the tumor (Chambers and Matrisian, 1997). The assumption that the host response to the tumor is immediately one of rejection of the invader tumor cells might not be entirely accurate; it might in fact be totally inaccurate. The presence of new malignant cells within the invaded tissue brings a new organization level, accommodation within the stroma and a steady nutrient supply and intense vascularization through the blood vessels and lymphatics. The transition from normal prostate epithelium to invasive carcinoma is preceded by or is concomitant with the activation of local host vascular channels and stromal fibroblasts. Stromal cell activation and recruitment by the tumor cell promotes malignant invasion, while neovascularization offers a portal for tumor dissemination (Liotta, Steeg, and Stettler-Stevenson, 1991; Dvorak *et al.*, 1995). Thus circulating endothelial precursor cells (CEP) migrate, elaborate degradative enzymes and traverse ECM barriers along a chemotropic gradient emanating from the tumor cells. Overcoming the androgen-independence of prostate tumors is clinically considered the most critical therapeutic endpoint for improving survival in patients with advanced metastatic prostate cancer. Most normal epithelial and endothelial cells undergo apoptosis when they become detached from the extracellular matrix (ECM), undergoing anoikis. In contrast, tumor cells are capable of evading anoikis and metastasizing to different distant organs (Douma *et al.*, 2004). However, pro-survival messages are required for migrating cells to resist the pro-apoptotic signals that take place during the disruption of integrin-mediated adhesion to ECM molecules (Mehlen and Puisieux, 2006). In parallel, transient ERK activation and augmentation of pro-survival pathways are potentially associated with cell migration. Is the biological repertoire of the epithelial and endothelial cells sufficient to account for the events associated with the process of anoikis during prostate cancer metastasis? While there is no clear answer to this question, what has become strikingly evident from existing knowledge is that molecules that induce anoikis in tumor epithelial and endothelial cells provide exciting new leads into effective therapeutic targeting, as well as being markers of prostate cancer progression and prediction of therapeutic efficacy/resistance (Garrison and Kyprianou, 2004). Thus the relevance of the anoikis phenomenon in prostate cancer metastasis and in the treatment of advanced disease is increasingly appreciated. It is critical to identify a predictive marker or a molecular signature before or during pre-clinical anti-tumor drug development to allow the selection of appropriate patients who are likely to respond, and the exclusion of patients who are unlikely to exhibit any clinical benefit.

The AR plays a pivotal role in the growth and survival of both normal prostate epithelium and abnormal neoplastic growth (Lamb, Weigel and Marcelli, 2001). The existing paradox is intriguing: while AR serves as a key regulator of prostatic function in the normal prostate epithelium (Prins, Birch, and Greene, 1991; Culig and Bartch, 2006), AR is converted into a promoter of uncontrolled growth in prostate tumors (Zhu and Kyprianou, 2008; Yuan and Balk, 2009; Harris *et al.*, 2009). Moreover, while loss of AR expression has been implicated as an underlying mechanism for the loss of androgen sensitivity of prostate tumors during development of androgen-independence and manifestation of clinical failure of prostate cancer patients to hormonal therapy (Gregory *et al.*, 2001b; Taplin, 2007; Knudsen and Scher, 2009), enhanced AR activity is involved in the early stages of prostate cancer (Litvinov, De Marzo and Isaacs, 2003; Heinlein and Chang, 2004). Recent molecular profiling revealed that progression from high-grade pre-malignant prostate hyperplasia to prostate carcinoma and metastasis is mediated by a selective down-regulation of AR target genes that inhibit proliferation and induce apoptosis (Hendricksen *et al.*, 2006). In a transgenic mouse model, overexpression of the wild-type AR in prostatic epithelial cells resulted in PIN but not prostate carcinoma (Stanbrough *et al.*, 2001). AR gene mutations in hormone-refractory prostate cancer that confer increased functional activity in the presence of ligands imply that resistance to androgen ablation is due to increased or altered sensitivity of the androgen signaling axis to other growth regulatory mechanisms (Zhu and Kyprianou, 2008; Knudsen and Scher, 2009). During prostate tumorigenesis the balance between the proliferation-inducing and apoptosis-enhancing functions of the AR can be disturbed not only by its mutations/inactivation but also via its interactions with signaling pathways.

The renowned pathologist Kerr and his colleagues initially described the morphologic features of apoptosis in response to hormone withdrawal from the rat ventral prostate (Kerr and Searle, 1973; Wyllie, Kerr and Currie, 1980; Sandford, Searle and Kerr, 1984). In contrast to the human prostate, however, the rat ventral prostate is composed of glandular epithelial cells with relatively sparse interglandular connective tissue. Castration in an adult male rat leads to rapid drop in serum testosterone levels to 10% of control within two hours, with a further drop to 1.2% within six hours. Intra-prostatic DHT level at 12–24 hours post castration is only 5% of that in the testis-intact control (Kyprianou and Isaacs, 1989). The rapid involution of the rat ventral prostate after castration is an active process initiated by removal of the inhibitory effects of androgen on prostatic cell death. The prostate gland undergoes a dramatic involution, by apoptotic deletion of the glandular epithelial cells, within two weeks of androgen deprivation (Kyprianou and Isaacs, 1988b). The death of the prostatic glandular epithelial cells induced by androgen ablation occurs as an active energy-dependent process involving a series of temporally discrete biochemical events resulting in the rapid programmed death of the subset of androgen-dependent cells within the rat ventral prostate (English, Santen and Isaacs, 1987). The cataclysmic drop in bioavailable androgens leads to dramatic changes in androgen receptor localization; the AR is no longer localized to the ventral prostatic nuclei 12 hours after castration (Kyprianou and Isaacs, 1988b; Zhang, Nan and Yu, 2002). By four days after castration, maximal DNA fragmentation is obtained with 15% of the total nuclear DNA detected as low molecular weight fragments. The dynamics of androgen-dependence are distinct: each day post castration, a subpopulation of androgen-dependent cells in rat ventral prostate fragmented all of their genomic DNA, as opposed to the whole population of cells fragmenting an increasing portion of their DNA daily. Associated with this apoptotic death, in those early days investigators were able to identify the enhanced induction of specific genes in the "programmed" prostate gland, including TGF-β expression (Kyprianou and Isaacs, 1989), early response proto-oncogene c-myc and c-fos (Buttyan *et al.*, 1988), and the TRPM-2 (testosterone-repressed message-2) gene (Montpetit, Lawless and Tenniswood, 1986). Apoptotic bodies start appearing 24 hours after castration, reach a dramatic peak at 48–72 hours, and are subsequently cleared by phagocytosis or secretion of digestive enzymes by neighboring cells (Buttyan *et al.*, 1999). Replacement of androgens leads to regrowth and restoration of prostate secretory activity (Isaacs *et al.*, 1992), further supporting the impact of androgen signaling on survival of the prostate glandular epithelial cells.

Regarding the role of stromal cells in castration-induced apoptosis of prostate glandular epithelial cells, elegant studies by Kurita and colleagues, using chimeric prostates with rat urogenital sinus mesenchyme plus bladder epithelium from wild-type or testicular feminization mutant (Tfm) mice, documented that AR-positive stromal cells are required for apoptosis induction in response to androgen withdrawal (Kurita *et al.*, 2001). A wealth of evidence generated in the 1990s made it absolutely clear that androgens stimulate production of vascular regulatory factors by stromal cells, including vascular endothelial growth factor (VEGF) and angiopoietins (Johansson *et al.*, 2005). The absence of these, post castration, leads initially to apoptotic loss of endothelial cells and vasoconstriction of the remaining blood vessels, followed by apoptosis of the glandular epithelial cells and ultimate regression of the gland (Joseph and Isaacs, 1997; Lekas *et al.*, 1997; Shabsigh *et al.*, 1998; Franck-Lissbrant *et al.*, 1998; Buttyan, Ghafar and Shabsigh, 2000).

The ligand TGF-β, secreted in a latent form and activated by proteolytic cleavage, regulates diverse biological activities including apoptosis, proliferation, cell motility, and angiogenesis in a number of human tumor epithelial and endothelial cells, as well as in smooth muscle cells (Coffey, Shipley and Moses, 1986; Gerdes *et al.*, 2004; Hjelmeland *et al.*, 2005). The breadth of cellular responses is dictated by the numerous genes and their encoded proteins regulated by TGF-β. Intracellular signaling by members of the TGF-β ligand superfamily proceeds via transmembrane heterotetrameric complexes comprising two types of serine threonine kinase receptors, type I (TβRI) and type II (TβRII), as rigorously documented by the pioneering studies from Massague's group (Siegel and Massague, 2003; Thomas and Massague, 2005). Upon ligand binding, TβRII receptor phosphorylates and activates TβRI receptor, which initiates the downstream signaling cascade by phosphorylating the receptor-regulated Smads (R-Smads) (Shi and Massague, 2003). Smad4 is the downstream effector essential for intracellular signal transduction by the R-Smads. Activation of R-Smads by their corresponding receptors results in heteromeric complex formation with Smad4 and upon nuclear translocation this complex results in context-dependent transcriptional regulation of TGF-β-targeted genes (Massague and Gomis, 2006). Smad7 is the antagonistic Smad that functions as a negative regulator of TGF-β signaling (Itoh *et al.*, 1998). Other signaling pathways also help to define the responses to TGF-β (Derynck and Zhang, 2003). Growing evidence suggests that in addition to Smads, TGF-β activates other effectors, and Smad-independent pathways also exist for TGF-β signaling (Derynck and Zhang, 2003; Zhu *et al.*, 2006). Deregulation of TGF-β signaling is critical to the pathogenesis of cancer due either to loss of expression or to mutational inactivation of its membrane receptors or intracellular effectors (Blobe, Schiemann and Lodish, 2000; Gold, 1999). TGF-β1 is closely linked to prostate tumor initiation and progression by eliciting both tumor suppressive (apoptotic, growth-inhibitory) (Guo and Kyprianou, 1999; Gerdes *et al.*, 2004; Zhu *et al.*, 2006) and oncogenic effects (EMT, angiogenesis) in metastatic cancer (Derynck, Akhurst and Balmain, 2001; Tuxhorn *et al.*, 2002c), either in conjunction with androgens or independently of the androgen axis (Sporn and Roberts, 1992; Bruckheimer and Kyprianou, 2001; Janda *et al.*, 2002).

# 3

# Apoptosis Pathways Signaling Execution of Cancer Cells

## 3.1 Cell Choices of Life and Death

As a unique mode of cell death, distinct from the non-specific necrotic death, apoptosis was originally defined by the father of medicine, Hippocrates, in his medical writings in the *MOXΛIKON*. He used the term ἀποπτωσεις (the plural of apoptoses) to describe the demarcation of bones that happens in a slow, gradual fashion (from *The Mochlicon*, Chapter 35) (Hippocrates, 1928). More than two millennia later scientists cleverly embraced the term "apoptosis" to define the programmed execution of cells via activation of their suicidal machinery, characteristically manifested at the pathological level as structurally dying cells with morphological intact organelles, meeting their fate in the ultimate degradation into "apoptotic" bodies, to be quickly phagocytosed.

Much of the current understanding of molecular mechanisms governing the apoptotic signaling pathways has come from a series of elegant genetic studies carried out by Robert Horvitz on a primitive nematode worm, *Caenorhabditis elegans* (*C. elegans*). These provided evidence of the evolution of the genetic programming of cell suicidal death from *C. elegans* to mammals (Ellis and Horvitz, 1986). Three main genes involved in apoptosis regulation during the development of *C. elegans* — CED-3, CED-4 and CED-9 — have mammalian counterparts, members of the BCL-2 family of apoptosis suppressors (CED-9 homolog); the cysteine proteases (caspases) that are apoptosis executioners (CED-3 homolog); and the apoptotic protease activating factor (APAF-1), death-promoting protein (CED-4 homolog) (Yuan *et al.*, 1993; Danial and Korsmeyer, 2004).

Cells undergoing apoptosis follow a well-recognized sequence of morphological patterns compared to non-apoptotic mode of cell death (Okada and

Mak, 2004). This morphological appearance is directly related to and can be explained by key intracellular molecular events, tightly coordinated to take place within the cell once the apoptotic trigger is received. In response to androgen ablation, the apoptotic process in the rat ventral prostate is manifested with gradual shrinking of cell size together with the formation of a condensed hyperchromatic nucleus (Kerr, Wyllie and Currie, 1972; Kyprianou and Isaacs, 1988a). This is accompanied by the breakdown of desmosomal junctions and loss of cell-surface elements causing the cells to lose contact with their neighbors (Kerr and Searle, 1973). The loss of mitochondrial membrane potential results in the release of calcium ions into the cytoplasm (Martikainen *et al.*, 1991). Calcium and magnesium-dependent endonucleases become active and cleave DNA at internucleosomal sites. Fragmentation of DNA prevents the transfer of potentially active genes from a dying cell to healthy nuclei in neighboring viable cells. There is subsequent condensation of the chromatin and segregation of the dying cells into membrane-bound vesicles. The mitochondria release their contents into the cytosoplasm, and apoptotic bodies with intact membranes and morphologically intact organelles are formed due to the active proteolysis by the family of caspases. This packaging prevents the leakage of cell contents into the extracellular space, which would otherwise stimulate an inflammatory reaction (Buttyan *et al.*, 1999). Engulfment of the apoptotic bodies is carried out by adjacent cells or by phagocytic cells of the immune system, which recognize ionic phospholipids and phosphatidylserine residues that characteristically appear on the apoptotic cell membrane during apoptosis. In the final stages of apoptosis the contents of the apoptotic bodies are rapidly degraded by lysosomes, and the entire apoptotic process is completed within a few hours. Yet the morphologically distinguishable cells, undergoing the execution phase of apoptosis at a given time before the intracellular cargo is degraded, are visible morphologically for perhaps less than one hour. All cytoskeletal proteins contribute to the changes that occur in apoptosis-like blebbing, shrinkage, and the formation of apoptotic bodies. Early transient polymerization and later depolymerization of F-actin are clearly manifested and associated with significant changes in the microfilament organization and cytoskeleton stability. DNA strand breaks due to activation of the endonuclease have been the original classic hallmarks of apoptosis detection (Kyprianou and Isaacs, 1988a; Kyprianou, English, and Isaacs, 1988). The traditionally accepted terminal deoxynucleotidyl transferase-mediated dUTP-biotin nick-end labeling (TUNEL) method involves labeling the 3-OH termini of DNA strand breaks with modified nucleotides. Considering that cell death via apoptotic elimination is a dynamic process and a rather transient phenomenon, capturing images of dying cells at a single cell level *in situ* might be a challenge, not only technically but also conceptually. While TUNEL has been a popularized and indeed an effective method applied by many investigators in diverse systems, the technique has characteristic advantages and disadvantages. Thus there are issues associated with the elaborate and rapid handling of tissues from selective detection of cells actually undergoing apoptosis, to elimination of false-positive and false-negative results for precise quantitative analysis and appropriate interpretation of the *in situ* data (Kyprianou, Bruckheimer and Guo, 2000). In our laboratory we routinely perform TUNEL staining analysis in various normal and tumor tissue specimens from the human and mouse prostate, and comparatively determine the rate of apoptotic index versus the proliferative index of the same cell populations (Berges *et al.*, 1995; McKenzie and Kyprianou, 2006). Figure 3.1 indicates a characteristic identification and

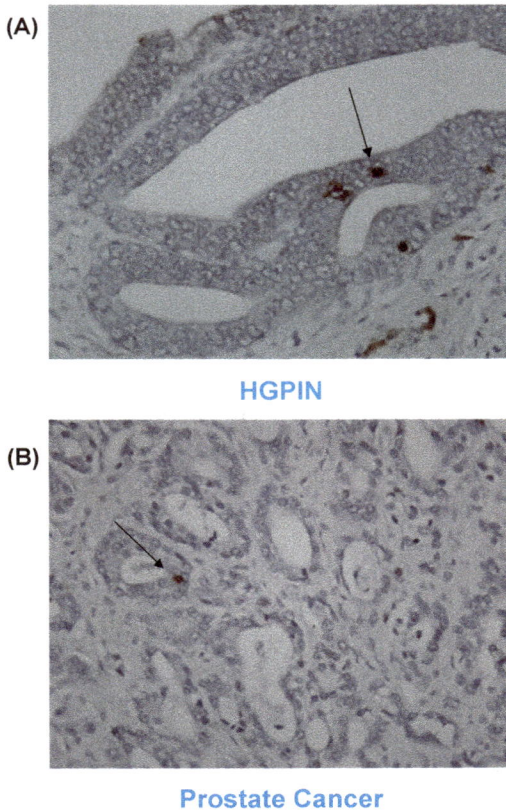

**HGPIN**

**Prostate Cancer**

**Figure 3.1** Detection of individual prostate tumor epithelial cells undergoing apoptosis, using the TUNEL assay in human prostate specimens. (A) High-grade prostatic intraepithelial neoplasia (HGPIN); (B) Well-differentiated adenocarcinoma of the prostate. Single cells undergoing apoptosis are detected as distinct brown TUNEL positive. Counterstaining is performed using H&E staining. Magnification 20x.

cellular distribution/localization of prostate tumor epithelial cells. These come from prostate cancer specimens from the transgenic adenocarcinoma mouse prostate (TRAMP) model of prostate tumorigenesis with increasing age, undergoing apoptosis as detected by the TUNEL assay at a given point in time.

### 3.1.1 *"Classic" apoptosis*

Apoptosis is triggered by signaling events involving a diverse array of protein networks, cellular organelles and macromolecular complexes that converge upon the activation of caspases. The executioner caspases are the proteases responsible, upon activation, for the cleavage of key proteins. This produces the characteristic apoptotic phenotypes of membrane blebbing, nuclear condensation, DNA fragmentation, and ultimately phagocytosis by the immune cells or adjacent macrophages (Wyllie, Kerr and Currie, 1980; Wolf and Green, 1999; Bruckheimer and Kyprianou, 2000). Mechanistically, apoptosis execution occurs via two distinct signaling pathways: the intrinsic and extrinsic apoptotic pathways are activated and propagated via intracellular effectors, after receiving the external death signals from the microenvironment (cytokine depletion, drug exposure, ionizing irradiation), ultimately leading to the death of the cell (Kyprianou, English and Isaacs, 1990). These are schematically described in Fig. 3.2.

- The *intrinsic pathway* is initiated by intercellular stress, lack of growth factors, or overwhelming DNA injury and subsequently targets the mitochondrial membrane (Hill, Adrain and Martin, 2003). Loss of mitochondrial membrane potential and increased membrane permeability leads to the release of cytochrome c into the cytosol (Du *et al.*, 2000; Verhagen *et al.*, 2000), that results in activation of apoptotic protease activating factor-1 (APAF-1) and caspase-9 recruitment. These proteins form a functional apoptosome that activates the effector caspase cascade, i.e. downstream activation of caspase-3 and -7, resulting in programmed cellular destruction (Fig. 3.2). The mitochondrial signaling pathway is tightly regulated through the pro-apoptotic and anti-apoptotic BCL-2 protein family members (Korsmeyer *et al.*, 2000). Proapoptotic members such as BAX, BID, and BAD allow for cytochrome c release and caspase cascade activation (DiPaola, Patel and Rafi, 2001). Once cytochrome c has been released into the cytosol, it binds to APAF-1 and stimulates APAF-1's oligomerization and the formation of an apoptosome. APAF-1 contains a caspase recruitment domain, CARD, a nucleotide-binding domain, and multiple WD-40 repeats. Alone, APAF-1 binds very weakly to dATP or ATP. Its binding with cytochrome c, however, increases its· affinity for dATP tenfold, which may occur as a result of a change in

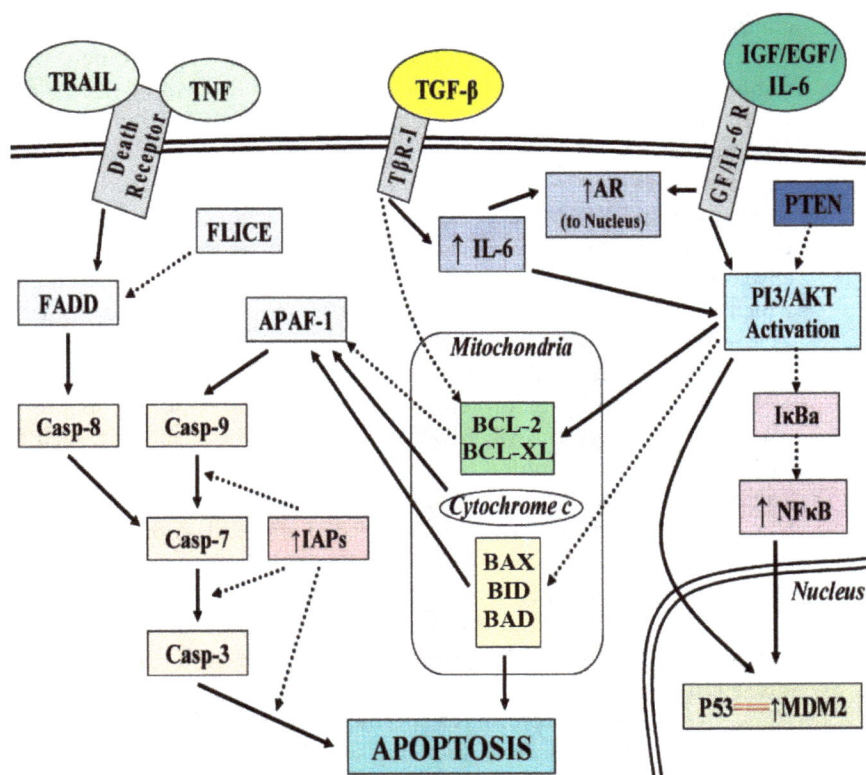

**Figure 3.2** Signaling cross-talk between the main intracellular effectors of apoptosis and survival pathways in prostate cancer cells "blend-in" during the intrinsic and extrinsic pathway of apoptosis signaling. The intrinsic pathway is initiated by intercellular stress, lack of growth factors, or DNA injury, targeting the mitochondrial membrane. Loss of mitochondrial membrane potential and increased membrane permeability leads to the release of cytochrome c into the cytosol, resulting in activation of APAF-1 (apoptotic protease activating factor) and caspase-9 recruitment. The subsequent formation of the apoptosome provides a caspase-activation platform leading to the downstream activation of caspase-3 and caspase-7, ultimately resulting in cellular destruction. The mitochondrial signaling pathway is by pro-apoptotic (BAX, BAD) and anti-apoptotic BCL-2 (B-cell lymphoma 2) protein family members (BCL-2, BCL-XL). The extrinsic pathway is initiated by the binding of FAS ligands or TRAIL at the extracellular domain of the cell surface death receptors associated with FADD. FADD contains an essential domain, death effector domain (DED), which allows FADD to bind to the initiator caspase, pro-caspase-8. Activated caspase-8 stimulates downstream effector caspases, such as caspase-7 and ultimately caspase-3, resulting in apoptosis. IAPs inhibit activation of effector caspases (activated in both the intrinsic and extrinsic signaling pathway), thus blocking cellular execution.

conformation subsequent to the binding of cytochrome c which leaves the nucleotide-binding site open; or the binding of cytochrome c may stabilize the complex formed between APAF-1 and dATP/ATP (Zou *et al.*, 1994). The APAF-1 oligomer subsequently induces the recruitment and accumulation of pro-caspase-9. Pro-caspase-9 undergoes auto cleavage and stimulates the activation of other effector caspases such as caspase-3 (Rodriguez and Lazebnik, 1999). Caspase-3 and -9 activate the enzyme I-CAD by cleaving it to produce CAD, which penetrates the nuclear envelope of the cell and proceeds to cleave cellular DNA. Cytochrome c, which resides in the mitochondrial inner membrane spaces, is released as a result of the interactions between members of the BCL-2 family in response to an apoptotic stimulus. As discussed above, a distinct group of the BCL-2 family of proteins supports induction of the apoptotic pathway such as BAX, BAK, BID and BAD; the other members, such as BCL-XL and BCL-W, inhibit the apoptotic pathway.

- The *extrinsic pathway* is initiated by the binding of apoptosis-inducing ligands such as FAS or TRAIL (TNF-related apoptosis-inducing ligand) at the extracellular domain of the cell surface death receptors associated with FAS-associated death domain (FADD) (Bodmer, Holler and Reynard, 2000; Peter and Krammer, 2003). The death domains will subsequently recruit adaptor proteins composed of both death domains and additional domains that can bind a pro-caspase (Kischkel *et al.*, 2000). Every death receptor has a specific adaptor and for FAS it is called FADD. In this case, the death-inducing signaling complex (DISC) constitutes FADD and pro-caspase-8 (Cartee *et al.*, 2003). FADD contains an essential domain known as the death effector domain (DED) which allows FADD to bind to the homologous sequences on the initiators, caspase and pro-caspase-8. The recruitment of pro-caspase-8 is accompanied by its oligomerization, which subsequently allows it to cleave itself, leading to its self-activation (Fig. 3.2). Activated caspase-8 stimulates downstream effector caspases such as caspase-3 and induces the degradation of cytosolic, cytoskeletal, nuclear proteins, and DNA (Kuwana *et al.*, 1998). In addition, caspase-8 is also responsible for cleaving BID, a pro-apoptotic member of the BCL-2 family that forms the principal link between the extrinsic and intrinsic pathways by stimulating the release of cytochrome c from the mitochondrion (Kuwana *et al.*, 1998).

Another similar but distinct pathway proceeds from the tumor necrosis factor receptor I (TNFRI) which is stimulated by its ligand TNF-α. The activation of TNFRI allows for the clustering of the TNF-R1-associated death domain-containing protein (TRADD), cellular proteins such as FADD, TNF-R-associated factor 2, and receptor-interacting protein (Liu *et al.*, 1996; Micheau and Tschopp, 2003). At this juncture, it is fundamental to note that two separate signaling complexes can occur. In one case, when TRADD finds FADD, FADD interacts directly with pro-caspase-8 leading to its oligomerization and eventually its autocleavage as shown in Fig. 3.2. In the other case, TRADD recruits a serine/threonine kinase known as the receptor-interacting protein (RIP) and the adaptor called TNF receptor associated factor-2 (TRAF2). RIP, a 74kDa protein constitutes three domains: an N-terminal kinase domain, an alpha-helical intermediate, and a C-terminal death domain (Kelliher *et al.*, 1998). RIP mediates the activation of nuclear factor kappa B (NF-κB) via the activation of the IκB kinase complex (IKK). TRAF2 will stimulate a MAP kinase signaling pathway that leads to the activation of Jun kinase and the transcription factor Jun, a member of the AP-1 complex. NF-κB and Jun N-terminal kinase (JNK) are believed to inhibit apoptosis in some cell types (Liu *et al.*, 1996).

Activated FADD then interacts with initiator enzymes of the caspase cascade, notably caspase-8, and mediates the formation of DISC (Muzio *et al.*, 1996). Formation of DISC further stimulates caspase-8 dimerization, resulting in further downstream effector caspase activation (caspase-3 and -7), and ultimately programmed cellular destruction through proteolytic cleavage of caspase substrates (Ashkenazi, 2002; Micheau and Tschopp, 2003). The primary endogenous inhibitor of the death receptor pathway is the FLICE-inhibitory protein (FLIP), consisting of two death effector domains (DEDs) and a C-terminus caspase-like domain, providing structural similarity to caspase-8 (Thome *et al.*, 1997; Valmiki and Ramos, 2009). Functionally, FLIP has a higher affinity for DISC compared to caspase-8, thus blocking caspase recruitment and subsequent activation (Scaffidi *et al.*, 1999).

The extrinsic pathway is directly connected with the mitochondrial pathway via caspase-8 which cleaves BID, a BH3 pro-apoptotic member of the BCL-2 family proteins (Korsmeyer *et al.*, 2000; Hu *et al.*, 2006a). The cleaved BID penetrates the mitochondrial membrane and triggers the release of cytochrome c. In addition, BID and the activated members of the BCL-2 family, such as BAX and BOK, permeabilize the mitochondria which further promotes the release of cytochrome c and SMAC/DIABLO (second mitochondria-derived activator of caspases) (Oliver *et al.*, 2005). The function of SMAC is to prevent the actions of anti-apoptotic factors, and therefore allows caspase-3 to cleave I-CAD, which ultimately leads to apoptosis. These activities are antagonized by the anti-apoptotic members, most notably BCL-2 and BCL-XL, inhibiting cytochrome c release and SMAC/DIABLO from the mitochondria. Moreover, as illustrated in Fig. 3.2, the inhibitors of apoptosis proteins (IAPs) directly inhibit caspase-3, -7 and -9 activation, thus reducing apoptosis (Schimmer, 2004).

### 3.1.2 *Anoikis*

Anoikis is a specific mode of apoptosis resulting from insufficient cell-matrix interaction (Frisch *et al.*, 1996; Frisch and Screaton, 2001). Normal cells utilize anoikis to prevent cell proliferation at inappropriate locations. Tumor cells, however, develop means to evade this process and therefore acquire the ability to detach and migrate into new sites that provide a nurturing microenvironment for their continuous proliferation (Valentijn, Zouq and Gilmore, 2004; Rennebeck, Martelli and Kyprianou, 2005). Whether anoikis reflects cellular changes within the tumor microenvironment, or structural and spatial rearrangements in the interaction between tumor epithelial and endothelial cells with the ECM, is still debated. Specific histological, molecular, and transcriptional events are commonly associated with prostate cancer progression, leading to the attractive and rather obvious possibility of EMT as a part of the metastatic process. The cytoskeletal rearrangements that tumor cells go through during the process of EMT and during the migration and invasion of blood vessels may determine the plasticity of the tumor cells and their sensitivity to anoikis (Fig. 3.3). Molecules that change in expression, distribution, and function during EMT, and that are causally involved in the process, include growth factors (such as TGF-$\beta$, insulin-like growth factor 1 (IGF-1)), transcription factors (Smads, *Snail*), cell adhesion molecules to the ECM (integrins, ECM proteins, galectins) and cell-to-cell adhesion molecules (E-cadherin) (Hiroashi, 1998), as well as extracellular proteases (matrix metalloproteinases (MMPs), caveolin) (Berrier *et al.*, 2002; Jiang and Muschel, 2002; Barrallo-Gimeno and Nieto, 2005; Handsley and Edwards, 2005). Three proteins emerge as leading molecular candidates functionally involved in the development of anoikis resistance during cancer progression: galectins, caveolin and tyrosine kinase (Trk)B (Califice *et al.*, 2004; Douma *al.*, 2004). While their specific role in EMT and RS formation awaits clarification, their direct targeting for therapeutic exploitation becomes an intriguing possibility. The development of EMT behavior by cancer cells however is only one hurdle in achieving metastatic "success", thus it is not expected that any EMT manifestation by prostate cancer cells will lead to immediate metastasis. Since the majority of human prostate tumors are vastly heterogeneous, one would expect only a few will show the molecular EMT-like signature towards an invasive front (as shown in Fig. 3.3).

### 3.2 Caspases: The Apoptosis Executioners in a Therapeutic Setting

The mechanism of apoptosis is remarkably conserved across species and is executed by a cascade of sequential activation of initiator (facilitator) and effector (executioner) caspases (Taylor, Cullen and Martin, 2008). Caspases represent a fascinating family of death facilitator and death executioner proteins.

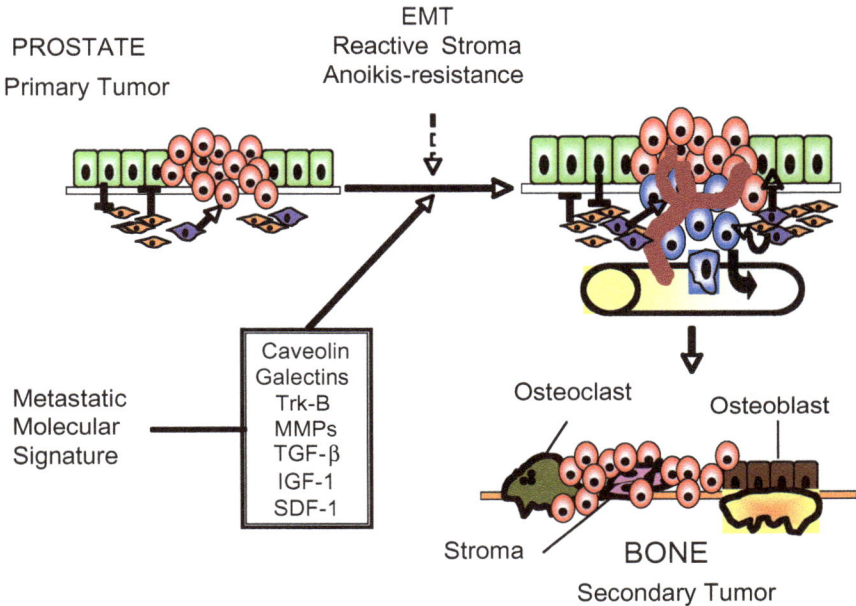

**Figure 3.3**  Significance of the tumor microenvironment in prostate cancer progression to metastasis to the bone. The reactive stroma promotes the anti-death behavior of tumor epithelial cells soon to become free but "homeless". Such homeless cancer epithelial cells can survive detachment from their "home ground" and resist anoikis by using autocrine or paracrine mechanisms from the surrounding stroma to suppress apoptosis, stimulate tissue invasion, and promote growth of new blood vessels to ensure a rigorous supply of oxygen. Prostate cancer epithelial cells can also undergo EMT at the primary site and upon detachment from the ECM, and after losing the connections with the neighboring cells, fail to undergo anoikis and gain a survival advantage thus entering the circulation. Continued resistance to anoikis facilitates dissemination into the circulation and colonization of the metastatic tumor cells in the bone by engaging active interactions with osteoclast and osteoblast populations. Molecular signature analysis of the metastatic phenotype of human prostate cancer cells identifies the key regulators of both processes functionally promoting metastasis (anoikis and EMT): TGF-β, galectins, caveolin, and MMPs.

This is a "close" family of cysteine proteases that are expressed as inactive pro-enzymes in normal, healthy cells; however, upon activation in response to an apoptosis trigger, the pro-enzyme is processed and the active heterotetramer is formed (Cohen *et al.*, 2007). Active caspases are orchestrated to selectively cleave intracellular target protein substrates at the carboxyl terminus of specific aspartate residues (Wolf and Green, 1999). There are twelve caspase isoforms in humans, but their precise and non-redundant roles in mediating apoptosis and innate immunity are only beginning to be fully elucidated. Within the growing caspase family, caspase-1, -2, -8, and -10 have been implicated in the initiation of apoptosis, whereas caspase-3, -6, -7, and -9 have been involved in the execution of apoptosis (Wang *et al.*, 2001; Cohen *et al.*, 2007). The vital significance of caspase pro-domains in the regulation of caspase activity is evidenced by the recognition of a family of adaptor IAPs, capable of binding the caspases and sequestering their proteolytic activity (Roy *et al.*, 1997). Extensive biochemical and structural analyses established that the inflammatory and initiator caspases are recruited to larger complexes that ultimately induce caspase dimerization, autoproteolysis and protease activation (Lakhani *et al.*, 2006; Jang *et al.*, 2007). The executioner caspases are translated as inactive dimers with at least two sites of processing. The specific pre-olytic event in the regulation of execu-tioner caspase activity remains the focus of several investigations, and much anticipation surrounds its precise functional definition at the cell level as well as *in vivo*. Genetic knockout studies demonstrated that in mouse embryonic fibroblasts derived from double knockout mice, cells deficient in both caspase-3 and caspase-7 are more resistant to a series of apoptotic stimuli compared to single knockout models, evidence that implicates a high degree of functional redundancy in these two executioner forms (Lakhani *et al.*, 2006).

The role the caspases play at both the commitment and execution phases of apoptosis, resulting in the controlled demolition of the cell, has recently led to a cascading exploitation and design of new therapeutic modalities for the treatment of advanced prostate cancer. In theory, caspases can be targeted at two levels: gene expression (gene therapy), and post-translationally (zymogen cleavage and caspase activation). Gene therapy directed at common down-stream effectors caspase-3 and caspase-7 should bypass intracellular checkpoint genes that limit apoptosis, making the cell unable to escape its death (Fig. 3.2). The downstream effector caspase, caspase-7, has been shown to be activated following apoptosis induction by several agents including lovastatin and okadaic acid in prostate cancer cells (Bowen *et al.*, 1999; Marcelli *et al.*, 1999). Of major significance is the observation that overexpression/direct activa-tion of caspase-7 induces apoptosis in BCL-2-overexpressing LNCaP prostate cancer cells (Marcelli *et al.*, 1999). Considering the significant role of BCL-2 expression in prostate cancer progression to androgen-independent disease (McDonnell *et al.*, 1992) and since caspase activation occurs downstream from the BCL-2 checkpoint (Kroemer, 1997), these findings have important clinical implications, as they suggest that caspase-7 activation can induce apoptosis in BCL-2-overexpressing prostate tumors. Based on the mechanistic outline of apoptotic pathways summarized in Fig. 3.2, one could promote caspase-7 and caspase-9 as potential therapeutic gene candidates for prostate cancer therapy. Moreover, the constitutive kinase activity identified with caspases has recently been considered as a novel target for gene therapy in advanced prostate cancer (Cornford *et al.*, 1999).

Chemical-induced dimerization (CID) of transfected caspase pro-forms may also prove an effective therapeutic strategy for androgen-independent prostatic tumors. The "proximity" model of transproteolysis for caspase activation supports the notion that the weak proteolytic activity of two pro-caspases being brought into close contact is sufficient for them to cleave to one another, form active enzymes, and thereby begin the proteolytic cascade (MacCorkle *et al.*, 1998). This model of chemically induced dimerization using modified pro-caspase molecules, termed artificial death switches, is being considered as a potential anti-cancer treatment (MacCorkle *et al.*, 1998). Moreover, direct activation of caspases, through CID within the human prostate smooth muscle cells, has been implicated as a potential effective treatment for benign prostatic hyperplasia and prostate cancer (Nikitina *et al.*, 2005). This provides a direct rationale for designing conditional alleles based on CID, and implementing chemically induced apoptosis as a novel and effective therapeutic strategy for prostate cancer.

The initial connection at the histopathological level between caspases as the death enzymes and tumor grade was reported a decade ago, at a time when caspase biochemical characterization was in its infancy. Thus a significant decrease in caspase-1 protein expression correlates with prostate tumor progression in prostate cancer patients (Winter, Rhee and Kyprianou, 2004). If prostate cancer cells require *de novo* synthesis of caspases, coordinating the timing of administration of chemotherapeutic agents that inhibit transcription or translation with those that induce apoptosis could prove critical in determining the effectiveness of therapy. The ability of certain agents to increase expression of caspases and subsequently prime cells for apoptosis may prove effective as adjuvant-based therapy. Studies in our laboratory demonstrated the potential of the prostate growth modulator TGF-$\beta$1 to induce prostate cancer cell apoptosis via up-regulation of caspase-1 expression (Guo and Kyprianou, 1999). Furthermore, apoptosis induction in androgen-independent prostate cancer cells in response to FTY720, a fungal-derived metabolite, requires caspase-3 activation (Wang *et al.*, 1999). Finally, recent evidence points to the mitochondrial respiratory chain as a functionally powerful therapeutic target for androgen-independent prostate cancer (Joshi *et al.*, 1999). These investigators demonstrated that apoptosis induction by a novel anti-prostate cancer compound, a fatty acid containing hydroxamic acid (MBD188), proceeds via an up-regulation of cytochrome c oxidase, a biphasic alteration in the mitochondrial membrane potential, and dramatic changes in the mitochondrial morphology prior to caspase activation. Considering that there is a disruption in the pro- and antioxidant balance during prostate cancer progression, these findings gain high clinical significance in targeting the electron transport chain of the mitochondria for the treatment of advanced prostate cancer.

## 3.3 The Mitochondrion: A Convenient Cell-Killing Platform

Mitochondria play a key role in the regulation and signaling of apoptosis pathways by providing a "centralized" platform for the execution of the genetic ordering for the cell to die (Newmeyer and Miller, 2003). A series of critical events in apoptotic signaling focus on the functional integrity of the mito-chondria, including the release of cytochrome c, calcium efflux, disruption of electron transport and energy metabolism, the generation of reactive oxygen species (ROS), and alterations in the intracellular redox potential (Kroemer *et al.*, 1997). Cytochrome c, which is normally present in the mitochondrial inner membrane space, is released into the cytosol, triggering apoptosis induction in response to various stimuli (Kuida *et al.*, 1998). Mitochondrial outer membrane permeabilization (MOMP) is often required for activation of caspase proteases. Various intermembrane space (IMS) proteins, such as cytochrome c, promote caspase activation following their mitochondrial release. As a consequence, mitochondrial outer membrane integrity is highly controlled, primarily through interactions between pro- and anti-apoptotic members of the BCL-2 protein family (Srivastava *et al.*, 1998). Following MOMP by pro-apoptotic BCL-2-associated X protein (BAX) or BCL-2 antagonist or killer (BAK), additional regulatory mechanisms govern the mitochondrial release of IMS proteins and caspase activity (Tait and Greene, 2010). Release of cytochrome c from the mitochondria and its subsequent binding to caspase-9 mediates the formation of the apoptosome by recruiting APAF-1 (Yoshida *et al.*, 1998; Kim *et al.*, 2008), resulting in the transactivation of pro-caspase-9 that triggers the sequential activation of the downstream apoptosis executioner caspase-3 (Fig. 3.2). Respiratory chain complexes I–IV generate the proton gradient over the mitochondrial inner membrane that drives ATP generation by ATP synthase. Executioner caspases (caspase-3 and caspase-7) enter the mitochon-drial intermembrane space following MOMP, disrupting key complexes. This caspase-directed action results in loss of transmembrane potential ($\Delta\Psi_m$) and ATP synthesis, and an increase in reactive ROS production. These effects of mitochondrial dysfunction contribute to the exposure of phosphatidylserine on the outer leaflet of the plasma membrane and its permeabilization, one of the characteristic features during apoptosis initiation (Tait and Green, 2010). MOMP is recognized as a point of no return for cell survival, irrespective of caspase activity following MOMP. As a consequence, intense investigative efforts are focused on addressing the mechanism by which the mitochondrial outer membrane is selectively breached, causing apoptosis.

## 3.4 Cell Surface Death Receptors and the FAS Ligand

FAS antigen/CD95 is a unique cell surface receptor protein that can initiate certain intracellular signaling pathways leading to apoptosis when engaged by its natural ligand, or when non-specifically activated by divalent antibodies against its internal domain (Micheau and Tschopp, 2003; Debatin and Krammer, 2004). The molecular ordering of the FAS apoptotic pathway has been fully dissected (Srinivasula *et al.*, 1996). The interaction of FAS with the FAS ligand FASL leads to the aggregation of FAS cytoplasmic domains (DD) and increases the affinity of the FAS DD for the FAS intracellular domains, the adaptor molecule FADD (MORT1) (Medema *et al.*, 1997). The evidence suggests that ceramide-mediated clustering is required for CD95-DISC formation (Grassme *et al.*, 2003). The end result of this FAS activation process is an unmasking of the proteolytic activity of caspase-8, an effector component of the apoptotic machinery (Peter and Krammer, 2003). The aggregation of caspase-8 contributes to the initiation of a protease cascade, probably by transproteolysis, that includes caspase-1 and caspase-3 and which ultimately leads to the irreversible cleavage of multiple pro-apoptotic targets. In the absence of caspase-8, death receptor recruitment of caspase-10 can initiate apoptosis (Kischkel *et al.*, 2001). Mitochondrial-derived factors, primarily cytochrome c, are essential for the activation of the most downstream members of this protease death cascade, including caspases-3, -7 and -9 (Fig. 3.2).

Death receptor-induced cell death is frequently encountered in prostate cancer in response to diverse stimuli (Rokthlin *et al.*, 2000; Aoudjit and Vuori, 2001; Guseva *et al.*, 2004; Thorpe *et al.*, 2008). The significance of the FAS antigen signaling cascade in prostate apoptosis has been demonstrated in the normal rat prostate following castration-induced apoptosis (de la Taille *et al.*, 1999). Growing evidence emerging from several *in vitro* studies implicates activation of the FAS-FAS ligand pathway in sensitizing androgen-independent human prostate cancer cells to undergo apoptosis in response to various chemotherapeutic agents (Costa-Pereira and Cotter, 1999; Rokthlin *et al.*, 2000; Garrison and Kyprianou, 2006; Thorpe *et al.*, 2008), as well as lymphocyte-mediated cell killing (Frost *et al.*, 1997). Thus if prostate epithelial cells harbor an intact FAS signaling pathway, sensitization of androgen-independent tumors to anti-FAS-induced apoptosis becomes an appealing therapeutic target, with potential clinical application in the effective treatment of advanced prostate cancer.

## 3.5 Meet the BCL-2 Family: Governors of Cell Survival and Death

BCL-2 was originally discovered as a proto-oncogene, via its association the t(14;18) chromosomal translocation characteristic of follicular lymphoma, its oncogenic potential confirmed a transgenic mouse model (Cleary and Sklar,

1985; Tsujimoto *et al.*, 1985; McDonnell *et al.*, 1989; Tsujimoto, 1989; Veis *et al.*, 1993). Historically its functional identity in the context of direct suppression of apoptosis rather than enhancement of cell proliferation emerged from a series of genetically-enforced expression studies elegantly conducted by the legendary Dr Korsmeyer (Hockenbery *et al.*, 1990; McDonnell and Korsmeyer, 1991). The BCL-2 family proteins include a heterogeneous group of both pro-apoptotic and anti-apoptotic molecules that exert their effect on mitochondrial function (Vaux, Cory and Adams, 1998; Di Paola *et al.*, 2001). All members of the BCL-2 gene family possess at least one BCL-2 homology domain and numerous BCL-2 family members have been identified based on these homologous regions. Functionally the domains allow hetero- and homodimerization between the BCL-2 family members, which regulate the apoptotic response (Hanada *et al.*, 1995; Kroemer, 1997; Perlman *et al.*, 1999). In a rheostat-controlled fashion, it is the ratio of pro-apoptotic to anti-apoptotic family members that ultimately determines tumor cell survival (Oltvai, Milliman and Korsmeyer, 1993; Nakayama *et al.*, 1994). Structurally, many BCL-2 family members contain a hydrophobic stretch of amino acids near their carboxyl terminus that anchors them in the outer mitochondrial membrane (Yang *et al.*, 1995). In contrast, other family members, such as BID, BCL-2 interacting mediator of cell death (BIM) and BAD, lack these membrane-anchoring domains, but target mito-chondria. Anti-apoptotic members, most notably BCL-2 and BCL-XL, inhibit the release of cytochrome c from the mitochondria, consequently inhibiting mitochondrial-induced apoptosis. Their action is heavily antagonized by pro-apoptotic members of the BCL-2 family such as BAX, BAD, and BID, allowing for mitochondrial cytochrome c release and caspase cascade activation (DiPaola, Patel and Rafi, 2001). Although it was initially considered that the interac-tions between the BCL-2 family members with respect to homodimerization versus heterodimerization determined the commitment to undergo cell death, individual BCL-2 family members have later been shown to independently regulate apoptosis, without requirement for heterodimerization (Cheng *et al.*, 1997; Korsmeye *et al.*, 2000). On activation, BAX and BAK undergo extensive conformational changes, leading to the mitochondrial targeting of BAX and the homo-oligomerization of BAK and BAX. Oligomerization of BAX or BAK is likely to be required for MOMP as mutants of either protein that fail to form oligomers are unable to cause MOMP. FRET-based analysis of BAX-mediated liposome permeabilization has provided compelling, real-time evidence for direct and dynamic interactions between truncated BID (tBID) and BAX, which precede BAX membrane insertion and liposome permeabilization (Biswas and Greene, 2002). This supports a model in which BAX (and by analogy BAK) acti-vation requires interaction with BH3-only proteins. Structural analysis of BAX in complex with a chemically stapled BCL-2 interacting mediator of cell death (BIM; also known as BCL2L11) BCL-2, homology regions (BH)3 domain peptide termed BIM SAHB (stabilized $\alpha$-helices of BCL-2 domains) revealed a somewhat unexpected interaction site. BIM SAHB does not bind in the BAX hydrophobic BH3-binding pocket (as occurs when the BID BH3 domain binds BAK) but, instead, binds on the opposite side of BAX. Mutations in BAX that inhibit BIM SAHB binding attenuate BAX-induced MOMP, supporting a functional relevance for this interaction during BAX activation (Biswas and Greene, 2002).

BIM is characterized to be the most potent pro-apoptotic protein that functions in the intrinsic apoptosis pathway, triggered via withdrawal of growth factor signals (Biswas and Greene, 2002; Gilley, Coffer and Ham, 2003; Wang, Gilmore and Streuli, 2004) or exposure to pro-apoptotic triggers (Enders *et al.*, 2003; Sunters *et al.*, 2003; Wang, Gilmore and Streuli, 2004). When epithelial cells detach from the ECM, BIM translocates to the mitochondria, where a vital interaction with BCL-XL further blocks its pro-survival function, while promoting BAX activation (Reginato *et al.*, 2003). Upon attachment to the ECM, integrin-mediated signals promote extracellular signal-regulated protein kinases (ERKs) and phosphatidylinositol 3-kinase (PI3K)/AKT-mediated phosphorylation of BIM, mediating its proteosome-dependent degradation. The onset of anoikis is followed by inhibition of the ERK and PI3K/AKT downstream signals and concomitant stimulation of BIM phosphorylation, which prevents BIM from degradation and intracellular accumulation, and subsequently promotes anoikis (Wang, Gilmore and Streuli, 2004; Giannoni *et al.*, 2008; Giannoni *et al.*, 2009).

In the normal prostate, BCL-2 expression is restricted to the basal cells of the glandular epithelium, which interestingly enough represent a cell population resistant to the effects of androgen deprivation-induced apoptotic effects (McDonnell *et al.*, 1992; Colombel *et al.*, 1993). Both *in vitro* and *in vivo* studies have established that BCL-2 and other anti-apoptotic members of its family are significantly up-regulated in aggressive prostate cancer phenotypes (McDonnell *et al.*, 1992; Colombel *et al.*, 1996; Kajiwara *et al.*, 1999). Immunohistochemistry-based investigations correlated BCL-2 overexpression contributes with the emergence of castration-resistant prostate cancer following androgen deprivation therapy (Apakama *et al.*, 1996; Colombel *et al.*, 1996). Thus, high abundance of BCL-2 may enable prostate cancer epithelial cells to survive in the absence of androgens, evidence that suggests that androgen ablation therapeutic regimes may actually select BCL-2-overexpressing apoptosis-resistant cells within a glandular cell population in the prostate epithelium (Tang and Porter, 1997). In patients with locally advanced prostate cancer, or metastatic prostate cancer receiving hormonal therapy, BCL-2 overexpression is an adverse prognostic indicator (Colombel *et al.*, 1996). Moreover, translational studies point to a predictive value of the apoptosis suppressors, as it is shown that over expression of BCL-2 and BCL-XL in human prostate tumors confers resistance to chemotherapy and radiation therapy (Gleave *et al.*, 1999; Scherr *et al.*, 1999; Szostak and Kyprianou, 2000; McCarty, 2004).

The interactions between the BCL-2 family members and the androgen signaling axis operating in prostate cancer cells, recognized as critical to the apoptosis-based outcomes of therapeutic responses, are intriguing in complexity as their role unfolds. Thus in androgen-responsive prostate cancer cells, androgens down-regulate expression of pro-apoptotic BCL-2 family members such as BAX (Coffey *et al.*, 2002). Increased BCL-2 and BCL-XL expression in androgen-independent tumors (Furuya *et al.*, 1996) is directly linked to the ability of prostate cancer cells to survive in an androgen-free environment (Kajiwara *et al.*, 1999), evidence highlighting not only the functional, but also the predictive significance of BCL-2/BCL-XL overexpression as one of the key regulators allowing for selection of androgen-independent recurrences in prolonged androgen ablation therapy. While recognizing that expression changes in the BCL-2 family cannot contribute to the emergence of therapeutic resistance alone, the dynamic cross-talk between this "powerful" family and other anti-apoptotic pathways influenced by exogenous ligand-receptor signaling must be considered in depth.

## 3.6 The Transcriptional Controllers

The NF-κB/Rel protein family of transcription factors regulates a multitude of immunologic and inflammatory responses, as well as cell growth, differentiation, and apoptosis (Kane *et al.*, 1999; Suh and Rabson, 2004). While the oncogenic properties of NF-κB include augmenting angiogenesis, invasion, and metastasis formation, the most important mechanism driving the carcinogenic effect of this transcription factor is its anti-apoptotic pathway towards cell survival (Chen, 2004; Suh and Rabson, 2004). In the majority of cell types, NF-κB is kept inactive through compartmentalization to the cytosol via binding with inhibitors of κBs (IκBs) (Fig. 3.2). In response to the appropriate stimuli, IκB is phosphorylated through interaction with IKK, allowing nuclear translocation of NF-κB and subsequent NF-κB-driven signal transduction (Chen, 2004; Karin, 2006). The anti-apoptotic effect of NF-κB is due to its ability to directly up-regulate BCL-2 and BCL-XL expression in prostate cancer cells, inhibiting mitochondrial apoptosis (Shukla and Gupta, 2004). However, thorough mechanistic dissection of the prostate apoptosis response-4 protein (PAR-4) pathway revealed that NF-κB inhibition is required for the pro-apoptotic effect of PAR-4 on the FAS ligand extrinsic apoptotic pathway to occur (Chakraborty *et al.*, 2001). Furthermore, NF-κB can up-regulate FLICE inhibitory protein (c-FLIP-L) expression in prostate cancer cells; such c-FLIP-L up-regulation directly interferes with recruitment of caspase-8 to FADD (Zhang *et al.*, 2004b). Down-regulation of c-FLIP-L appears to restore functional sensitivity to FAS-mediated apoptosis in aggressive prostate cancer cells (Hyer *et al.*, 2002). Downstream effector caspase activation is not immune to the inhibitory effects of NF-κB either, as it has been shown to up-regulate several IAPs that directly inhibit caspases-3, -7 and/or -9 (McEleny, Watson and Fitzpatrick, 2001).

Constitutive activation of NF-κB occurs in low levels in androgen-dependent prostate tumors, but it reaches higher levels in androgen-independent cancer cells and in highly aggressive prostate tumors and metastatic lesions (Ismail *et al.*, 2004; Karin, 2006). In human prostatectomy specimens, elevated NF-κB immunoreactivity correlated directly with advanced tumor stage, tumor grade, and tumor recurrence (Ross *et al.*, 2004). In lymph node metastasis, nuclear localization of NF-κB is significantly greater in lymph nodes containing tumor cells when compared to local tumor controls (Ismail *et al.*, 2004). Interestingly, high NF-κB activation is also detected in tumor-surrounding lymphocytes, suggesting possible cross-talk between prostate cancer cells and surrounding lymphocytes, leading to oncogenically favorable release of paracrine prostate cancer mediators such as IL-6 and TNF-α (Viatour *et al.*, 2005). This up-regulation of NF-κB found in aggressive prostate cancers correlates with resistance to both chemotherapy and radiation therapy (Sumitomo *et al.*, 1999; Suh and Rabson, 2004). While there appears to be a direct correlation between increased NF-κB activity and androgen independence, the functional interaction between the NF-κB pathway and the AR signaling appears to be pleomorphic (Coffey *et al.*, 2002) and merits further elucidation.

## 3.7 The p53 Tumor Suppressor

The p53 tumor suppressor gene regulates both cell cycle and apoptosis in response to numerous cellular stresses such as DNA injury, hypoxia, free radical injury, and damage to the mitotic machinery (Hernandez *et al.*, 2003). The oncolytic responsibility of the p53 gene product is to invoke cell cycle arrest and stimulate apoptosis in cells that have acquired overwhelming genetic aberrations to avoid mutation propagation (Stellmach *et al.*, 1996). The action of p53 is inhibited by MDM2, through binding to the p53 gene product and relegating it to ubiquitylation and proteosomal degradation (McCarty, 2004). Mechanistically the ability of p53 to induce apoptosis in tumor cells results from its up-regulation of BAX, leading to mitochondrial-driven apoptosis, potentially by blocking the AKT survival signaling (Hernandez *et al.*, 2003) (Fig. 3.2). Furthermore, analysis of specific p53 mutations revealed that altered p53 expression can also adversely affect FAS-mediated apoptosis (Gurova *et al.*, 2005).

Loss or mutation of p53 has been identified in over 10,000 different types of analyzed tumors and mutations of this tumor suppressor gene are found in 45–50% of all human cancers (Soussi, Dehouche and Beroud, 2000). In prostate cancer, while mutations in p53 are uncommon in early, well-differentiated disease, mutations become abundant in both metastatic disease and hormone-independent tumors (Navone *et al.*, 1993). Moreover, up-regulation of MDM2 expression has been found in up to 40% of prostatectomy specimens and correlated with advanced disease. In hormone-responsive prostate cancer cells, loss of wild-type p53 leads to development of a hormone-resistant phenotype, thus implicating altered p53 function in the development of hormone refractory disease (Scherr *et al.*, 1999; Burchardt *et al.*, 2001). Significantly enough, altered p53 can impact the chemotherapeutic response, with most mutations leading to resistance, while certain select mutations lead to increased sensitivity to specific agents such as paclitaxel (DiPaola, Patel and Rafi, 2001). There is growing support for a direct correlation between altered p53 expression and prostate cancer resistance to radiotherapy (Pisters *et al.*, 2004), while resolution of functional p53 status with specific p53 mutants restores the apoptotic signaling and ultimately therapeutic sensitivity in experimental models of prostate cancer (Gudkov, 2003; Hernandez *et al.*, 2003).

## 3.8 PTEN/PI3K/AKT: The Downstream Intracellular Players

Alterations in phosphoinositide 3-kinase (PI3K)/mammalian target of rapamycin (mTOR) signaling are perhaps the most frequent events observed in solid tumors. Phosphatase and tensin homolog deleted on chromosome 10 (PTEN) is a highly conserved tumor suppressor gene that induces cellular apoptosis through its modulation of the PI3K/AKT signal transduction pathway. Specifically, PTEN inhibits phosphorylation of AKT, which is necessary for its activation and targeting of its many downstream effectors (Wang *et al.*, 2003). The PI3K/mTOR pathway can be activated by overproduction of growth factors or chemokines, loss of INPP4B or PTEN expression, or by mutations in growth factor receptors, Ras, PTEN, or PI3K itself (Fig. 3.2). Activation of this multidirectional pathway contributes to cell growth, survival and motility. Loss of PTEN, a common event in treatment-resistant and poorly differentiated prostate cancers, leads to constitutive activation of the PI3K/AKT pathway and subsequent apoptotic resistance (Davies *et al.*, 1999). Restoration of PTEN activity in PTEN-deficient prostate cancer cell lines increases sensitivity to FADD-mediated caspase-8-driven apoptosis, and facilitates BID cleavage allowing for cytochrome c release and subsequent mitochondrial apoptosis (Yuan and Whang, 2002). AKTs are activated by second messengers via phosphatidylinositol 3-kinases (PI3Ks), a phosphorylation event that is counterbalanced by the activity of PTEN phosphatases (Stern, 2004). In prostate cancer, AKT phosphorylation can occur constitutively through loss of PTEN activity, or be stimulated and up-regulated in PTEN-positive tumors through autocrine and paracrine cell membrane receptor–ligand interactions (Pfeil *et al.*, 2004). The ability of phosphorylated AKT to inhibit prostate cancer cellular apoptosis is the ultimate product of dynamic cross-talk between this effector and multiple other anti-apoptotic pathways, as shown in Fig. 3.2. Activated AKT can activate MDM2, leading to proteolysis of p53 and subsequent inhibition of p53-mediated apoptosis, with promotion of cell cycle progression (Gao *et al.*, 2003; Stern, 2004). Activated AKT also inactivates BAD and caspase-9, allowing for BCL-2 release and inhibition of mitochondrial apoptosis (Ghosh *et al.*, 2003; Wang *et al.*, 2003). Furthermore, up-regulation of PI3K/AKT activity leads to phosphorylation of IκB, nuclear translocation of NF-κB, and subsequent NF-κB-driven suppression of apoptosis (Stern, 2004; Karin, 2006). Activation of the survival signaling via phosphorylated AKT induces AR phosphorylation, leading to inhibition of androgen deprivation-induced apoptosis (Lin *et al.*, 2001; Ghosh *et al.*, 2003).

An expanding body of evidence defines both PTEN inactivation and AKT phosphorylation as hallmarks of aggressive prostate cancer. While PTEN inactivation is present in only 10–15% of primary prostate tumors, PTEN loss is detected in 30–50% of hormone-refractory tumors, as well as 60% of xenograft models derived from metastatic prostate cancer cell lines (Wang *et al.*, 2003). Furthermore, loss of PTEN correlates with aggressive local disease (T3b-T4 tumors) and Gleason score higher than 6 (McMenamin *et al.*, 1999). AKT phosphorylation (indicative of survival pathway activation), emerges as a predictor of aggressive disease, and an independent one at that. High levels of AKT phosphorylation are exclusively associated with prostatic adenocarcinoma, compared to benign tissues (Ayala *et al.*, 2004). Moreover, AKT phosphorylation correlates with Gleason score, and, in poorly differentiated tumors (Gleason 8–10), the strong presence of phosphorylated AKT is observed in over 90% of specimens examined (Malik *et al.*, 2002; Liao *et al.*, 2003). In prostate cancer specimens with Gleason scores of 5–6, a notoriously difficult patient population for predicting prognosis, elevated AKT phosphorylation proved to be a more effective prognostic indicator of recurrence than both mitotic index and Gleason score (Kreisberg *et al.*, 2004). Unsurprisingly, loss of PTEN and AKT phosphorylation are associated with resistance to chemotherapy and progression of castration-resistant prostate cancer (Ghosh *et al.*, 2003; McKenzie and Kyprianou, 2006). Strategies targeting AKT phosphorylation or restoring PTEN activity can cause profound apoptosis and restore sensitivity to chemotherapy *in vitro* and in xenograft models (Yuan and Whang, 2002; Shaw *et al.*, 2004). Rapamycin analogs have already been shown to have anti-tumor efficacy in some tumor types (Bjornsti and Houghton, 2004a). The therapeutic significance of mTOR targeting is strong for the treatment of patients with advanced prostate cancer, given the high prevalence of activation of the PI3K/AKT pathway due largely to the loss of expression/function of PTEN and the association of this pathway with adverse pathologic features, recurrence after radical prostatectomy, and systemic treatment resistance (Whang *et al.*, 1998; McMenamin *et al.*, 1999; Thomas *et al.*, 2004). Built on this solid evidence therefore, a rapidly expanding generation of new PI3K, AKT, and mTOR inhibitors has shown significant promise pre-clinically and is now in clinical trials with prostate cancer patients. Pre-clinical studies have shown an ability of target of rapamycin (TORC1) inhibitors to revert prostatic intraepithelial neoplasia and reduce tumor volume and growth/proliferation, particularly in tumors with activated AKT or that lack PTEN (Majumder *et al.*, 2004; Wu *et al.*, 2005). In a recently published clinical study, a short course of the mammalian target of rapamycin (TORC1) inhibitor was administered to men with intermediate- to high-risk localized prostate cancer before radical prostatectomy. It was found that although the activity of the downstream mammalian target of rapamycin target S6 was targeted as expected, no significant changes in tumor apoptosis or cell proliferation were detected in the majority of prostate tumors in response to rapamycin, and there was no detectable increase in AKT activation (Armstrong *et al.*, 2010). Although an anti-angiogenic or novel anti-tumor mechanism of action for rapamycin cannot be excluded, these findings suggest that single-agent TORC1 inhibition may be insufficient to have an effect on prostate tumor cell survival and apoptosis.

## 3.9 The Antagonists of Death: Inhibitors of Apoptosis Proteins (IAPs)

Recently, a new family of apoptosis inhibitors has been identified and appears to have a role in prostate cancer treatment resistance. The inhibitors of apoptosis proteins (IAPs) are a group of caspase inhibitors that directly inhibit the effector caspases-3, -7 and -9, resulting in decreased cellular apoptosis (Schimmer, 2004) (Fig. 3.2). Currently, eight human IAPs have been identified, with the most studied being X-linked inhibitor of apoptosis protein (XIAP), inhibitor of apoptosis protein 1 (IAP1), inhibitor of apoptosis protein 2 (IAP2) and survivin (Krajewska *et al.*, 2003). While all appear capable of inhibiting effector caspases, IAP1 and IAP2 can up-regulate NF-κB expression, pointing to a possible positive feedback loop between these two pathways (McEleny, Watson and Fitzpatrick, 2001). High expression of the four IAPs is found in both animal models of prostate cancer and human prostatectomy specimens from cancer patients, and this elevation appears to be present early in prostate cancer development (Krajewska *et al.*, 2003). The ability of IAPs to inhibit apoptosis in response to multiple chemotherapeutic agents has been established in several tumor models (Debatin and Krammer, 2004), although the significance of IAPs in prostate cancer therapeutic resistance is an area of recently "active" investigations. Indeed recent evidence suggests that XIAP inhibition enhances chemotherapy sensitivity in otherwise resistant prostate cancer cell lines (Amantana *et al.*, 2004). In a small clinical study, a dramatic up-regulation of IAP1 and IAP2 expression was found in patients receiving neoadjuvant androgen ablation, suggesting a potential role of IAPs in androgen independence (McEleny, Watson and Fitzpatrick, 2001). As the role of IAPs in prostate cancer treatment resistance continues to be revealed, manipulation of IAP pathways is quickly becoming a valuable way of circumventing the apoptotic resistance conferred by the upstream intracellular apoptotic escape mechanisms such as AKT and BCL-2. Survivin is another apoptosis suppressor (via its ability to promote cell survival) that has been validated as a cancer therapeutic target (Altieri, 2003).

### 3.10 Apoptosis Signaling in the Endoplasmic Reticulum: A Death Platform for Stress

Three ER-resident transmembrane proteins (IRE1, ATF6 and PERK) serve as sensors for ER stress (Shen *et al.*, 2002). The endoplasmic reticulum stress response (ESR) carried out by the individual sensor molecules provides a protective feature to ensure survival until the stimulus is eliminated (Kaufman, 1999; Hollien and Weissman, 2006). For example, PERK and ATF6 null MEFs die much more readily from ER stress induced by thapsigargin and tunicamycin (Harding *et al.*, 2000; Yamamoto *et al.*, 2007). When the ER stress perturbation is too advanced and massive, however, apoptosis can result (Ferri and Kroemer, 2001; Breckenridge *et al.*, 2003; Novoa *et al.*, 2003; Rao, Ellerby and Bredesen, 2004). While the precise events driving this effect remain unclear, one line of thought suggests that either a prolonged or a sufficiently great induction of one or more ER stress sensor proteins triggers an irreversible apoptotic cascade through induction of the ATF/XBP transcription factors (Cox, Shamu and Watter, 1993; Yoshida *et al.*, 2001; Jarosch, Lenk and Sommer, 2003). CHOP/Gadd153 is an ER stress target that is transcriptionally up-regulated by ATF6 (Scheuner *et al.*, 2001) and encodes for a transcription factor with apoptotic activity (Marciniak *et al.*, 2004). Its pro-apoptotic function can be blocked by GRP78 overexpression, thus placing CHOP apoptotic pathways as downstream from the ER (Wang *et al.*, 1996; Marciniak *et al.*, 2004). Mechanistically, CHOP may control apoptosis via both intrinsic and extrinsic pathways by down-regulating BCL-2, while upregulating DR5 expression (McCullough *et al.*, 2001; Yamaguchi and Wang, 2004).

In addition to translational repression, PERK has also been shown to have apoptotic activities mediated through eIF2α activity. The PERK-associated eIF2α has two phosphatase cofactors, Gadd34 and CReP, which can mediate apoptotic signaling downstream from the ER. Gadd34$^{-/-}$ cells exhibit constant eIF2α phosphorylation in the presence of an ER stress, resulting in fewer mis-folded proteins in the ER lumen (Kojimak *et al.*, 2003; Novoa *et al.*, 2003; Marciniak *et al.*, 2004). Furthermore elegant genetic studies showed that the Gadd34$^{-/-}$ mice are resistant to toxicity induced upon tunicamycin treatment (Marciniack *et al.*, 2004), while CReP knockdown provides protection from ER stress (Jousse *et al.*, 2003). In addition, pharmacological eIF2α inhibition by the agent salubrinal provides protection from ER stress-mediated apoptosis (Tsai, Ye and Rapoport, 2002; Boyce *et al.*, 2005). Another potential mechanism by which ER stress may induce apoptosis is dependent on IRE1, but does not rely on its endoribonuclease activity. IRE1 can interact with the TRAF2 adaptor protein to recruit ASK1, a pro-apoptotic kinase. This interaction leads to the formation of an IRE1/TRAF2/ASK1 complex that activates JNK (Nishitoh *et al.*, 2002). As ASK1$^{-/-}$ cells are partially resistant to apoptosis, this ternary complex may potentially regulate ESR-induced cell death (Nishitoh *et al.*, 1998). Moreover, BAX and BAK can initiate apoptosis by virtue of their localization in the endoplasmic reticulum (Zong *et al.*, 2003), a topologically powerful co-distribution that enables their physical association with IRE1 in the ER (as a stage) (Lee *et al.*, 2002a) and positively regulates their pro-apoptotic activity (Bertolotti *et al.*, 2000; Hetz *et al.*, 2006).

Due to their central role in dictating cellular life-and-death decisions, it is no surprise that the BCL-2 family can functionally control ESR-induced apoptosis. The regulation of calcium homeostasis by the ER provides a direct link to BCL-2 activity. ER Ca$^{2+}$ dynamics is a principal avenue whereby BCL-2 phosphorylation alters susceptibility to apoptosis. When BCL-2 is phosphorylated, Ca$^{2+}$ discharge from the ER is increased, with a secondary increase in mitochondrial Ca$^{2+}$ uptake, thus enabling a mechanism through which post-translational modification of BCL-2 inhibits its anti-apoptotic activity (Basu and Halder, 1998; Bassik *et al.*, 2004). Moreover, BAX and BAK overexpression is sufficient to enhance Ca$^{2+}$ release into the mitochondria, leading to permeabilization and increased cytochrome c release. Furthermore, MEFs lacking BAX and BAK exhibit notable resistance to ER stress-induced apoptosis, a finding which is substantiated by BAX$^{-/-}$/BAK$^{-/-}$ cells exhibiting reduced Ca$^{2+}$ release from the ER in response to arachidonic acid and oxidative stress (Wei *et al.*, 2001; Zong *et al.*, 2003; Scorrano *et al.*, 2003). Other studies involving ER stress stimuli established that the presence of BAX and BAK at both the ER and mitochondria is indispensable for normal apoptosis execution (Scorrano *et al.*, 2003) and proceeds via an APAF-1-independent pathway (Rao *et al.*, 2002). As in the other apoptotic signaling pathways, NF-κB has recently found its connection to the ER stress-induced apoptotic cell death. In MCF-7 cells, NF-κB regulation suppresses ER stress-induced apoptosis (Jiang *et al.*, 2003). Interestingly, IRE1-dependent NF-κB regulation has also been associated with TRAF2 down-regulation, up-regulation of TNF-α expression and signaling through TNF receptor 1. Therefore, NF-κB may enable a cross-talk between ER-induced apoptosis and the death receptor mechanism (Plember and Wolf, 1999; Rao, Ellerby and Bredesen, 2004; Hu *et al.*, 2006b).

The class of initiator caspases activated is a hallmark of the signaling pathway that the cell engages towards apoptosis induction. For instance, caspase-9 is an initiator for the intrinsic pathway, while caspases-8 and -10 are activated upon extrinsic stimuli. Two lines of evidence have suggested that caspase-12 is the initiator caspase for the ESR apoptotic response. First, caspase-12 is found in the ER membrane and can directly interact with IRE1 and the adaptor protein TRAF2 (Yoneda *et al.*, 2001). Second, caspase-12 is selectively processed through ER stress and not by other apoptotic stimuli (Nakagawa *et al.*, 2000). The involvement of caspase-12 in ESR-induced apoptosis remains controversial due, in part, to the absence of a human homolog (caspase-12 is expressed in rodents and not in primates) (Schroder and Kaufman, 2005). The recent association of human caspase-4 as the caspase-12 human homolog, as well as its similar localization in relation to the ER membrane, points to a functional overlap between the two caspases in ER stress-induced apoptosis (Hitomi *et al.*, 2004). However, the role of the less recognized caspases, caspase-4 and -12, in the apoptotic ESR has not been supported in cells lacking these genes (Obeng and Boise, 2005); this study demonstrated that caspase-9 was the sole initiator caspase responsible for apoptosis induction in response to ER stress stimuli. A potential interactive scenario can be considered involving the recruitment of all three apoptosis-signaling mechanisms — the mitochondrial, death receptor and ER stress apoptotic pathways — by the cancer cell towards its apoptotic execution when triggered by various ER stress stimuli (Kischkel *et al.*, 1995; Scorrano *et al.*, 2003; Scheel-Toellner *et al.*, 2004; Rao, Ellerby and Bredesen, 2004).

## 3.11 The Tumor Microenvironment: Extracellular Forces Control Intracellular Death Outcomes

Recently, the field of tumor biology has embraced a crucial paradigm shift implicating not only the intracellular pathophysiology inherent to carcinoma cells, but also the critical role that the tumor microenvironment plays in developing aggressive cancer phenotypes (Lawler *et al.*, 2001). Indeed, primary tumors frequently express genes whose protein products dramatically alter the microenvironment, such as extracellular matrix proteases, glycosylases, pro-angiogenic factors, regulators of cell adhesion, and mediators of inflammation (Camps *et al.*, 1990; Byrne, Leung and Neal, 1996; Burfeind *et al.*, 1996; Jung *et al.*, 1998; Ismail *et al.*, 2004). It has become apparent from accumulating evidence in experimental models (Bhowmick *et al.*, 2004; Carver and Pandolfi, 2006), that the cross-talk that exists between prostate epithelial tumor cells and their surrounding stromal and endothelial partners (Boddy *et al.*, 2005; Sakamoto, Ryan and Kyprianou, 2008), as well as the localized inflammatory cells attracted to the neoplastic region, is a driving force towards androgen-independence, bone metastasis, and subsequent treatment-resistant recurrence (Arnold and Isaacs, 2002; Bogdanos *et al.*, 2003; Chung *et al.*, 2005; Rennebeck, Martelli and Kyprianou, 2005).

### 3.11.1 *Role of hypoxia*

The ability of the tumor microenvironment to alter the apoptotic outcomes of prostate cancer cells is exemplified by recent advances in the understanding of tumor hypoxia. Solid organ tumors, including prostate adenocarcinoma, can outgrow their own blood supply. Coupled with their innate propensity for disordered neovascularization, this creates an intratumoral environment which often has a severely diminished oxygen tension. Hypoxia induces activation of hypoxia-inducible factor 1-α (HIF-1α), a nuclear transcriptional factor that is established under conditions of hypoxia, and which promotes the transcription and translation of several target genes that are critical for angiogenesis invasion and metastasis (Vaupel and Mayer, 2007). HIF-1α protein has been found to be overexpressed in most human malignancies, including prostate cancer (Rudolfsson and Bergh, 2009). Poor oxygenation and increased HIF-1 is the prime angiogenesis promoter both in normal and malignant prostate, by increasing angiogenic factors such as VEGF and its receptors (Hochachka, Rupert and Goldenberg, 2002; Rudolfsson and Bergh, 2009). Consequently, tumor hypoxia correlates significantly with prostate cancer stage, its aggressiveness, androgen independence, and therapeutic resistance (Movsas *et al.*, 2000; Ghatar *et al.*, 2003) and mutations in the oxygen-dependent degradation domain of HIF-1α are associated with increased risk for prostate cancer (Orr-Urtreger *et al.*, 2007). This hypoxia-driven aggressive behavior can, at least in part, be attributed to enhanced apoptotic resistance in hypoxic prostate tumor cells due to suppression of p53 activity and AKT up-regulation (Skinne *et al.*, 2004; Liu *et al.*, 2005b). Taken together, these observations led to an exploitative targeting of hypoxia towards enforcing prostate tumor regression (Pouyssegur, Dayan and Mazure, 2006). Along with tumor hypoxia, the paracrine and subsequent autocrine release of both growth factors and cytokines impacts apoptotic sensitivity and, ultimately, tumor aggressiveness. These exogenous proteins provide yet another set of critical signaling pathways in prostate cancer cell apoptotic evasion, in the form of anoikis resistance towards endothelial cell invasion and metastasis promotion (Fig. 3.3). That being said, there is evidence to indicate that androgen ablation leads to temporary prostate tissue hypoxia, triggering temporal apoptosis of the glandular epithelial cells (Rudolfsson and Bergh, 2009).

### 3.11.2 *The key growth factors*

Mechanistic dissection of the pathways leading to the emergence of castration-resistant prostate cancer identified the dynamic contribution of an array of growth factors, in addition to the androgen signaling axis. While a full understanding of the pro-survival characteristics of these growth factor pathways is still evolving, the impact of growth factors such as epidermal growth factor

(EGF), insulin-like growth factor 1 (IGF-1), and transforming growth factor-β (TGF-β) can be appreciated by the rigorous pharmacological development of targeted therapies impairing their signal transduction. As the medical and scientific community enthusiastically witnessed the development of the therapeutically promising tyrosine kinase inhibitor IRESSA, the role of the EGF system in apoptosis evasion and prostate cancer progression has been exposed. EGF can be secreted in a paracrine, then in an autocrine manner in prostate tumors (Mimeault, Pommery and Henichart, 2003). Up-regulation of its membrane receptor in invasive prostate carcinoma cells has also been well characterized (Hernes *et al.*, 2004; Shuch *et al.*, 2004). While up-regulation of this pathway has been associated with increased cellular proliferation and invasion (Mimeault, Pommery, and Henichart, 2003), its role in prostate cancer therapeutic resistance and androgen-independent status is typically attributed to its ability to protect prostate cancer cells from apoptosis. Indeed, the EGF-EGFR system can activate the survival pathway of PI3K leading to AKT phosphorylation and subsequent inhibition of pro-apoptotic BAD (Fig. 3.2). Interestingly, while disruption of the EGF-EGFR pathway results in profound prostate cancer cell apoptosis (Harper *et al.*, 2002; Farhana *et al.*, 2004), the enhanced apoptotic sensitivity can be only partially explained by down-regulation of the PI3K/AKT pathway. Mechanistic profiling of this pathway revealed the ability of the EGF-EGFR system to rescue prostate cancer cells from the pro-apoptotic effects of PI3K/AKT inhibition (Torring *et al.*, 2003). The ability of the EGF-EGFR system to provide apoptotic evasion in a PI3K/AKT-independent manner can be attributed, at least in part, to its effects on androgen receptor signaling. While the ability of epidermal growth factor to induce AR transcriptional activity alone has been a topic of debate (Orio *et al.*, 2002; Mellinghoff *et al.*, 2004), the ability of this system to coactivate the AR, sensitize it to the low levels of androgen characteristic of hormone ablation therapy, and synergize androgenic stimulation of AR transcriptional activity has been firmly established (Orio *et al.*, 2002; Gregory *et al.*, 2004; Zhu and Kyprianou, 2010). Such features may characteristically predict the clinical outcome of impairing the EGF-EGFR signaling and its consequences on prostate cancer progression and therapeutic resistance.

Like the EGF-EGFR pathway, the insulin-like growth factor (IGF) axis has proven to be a critical player in the progression of prostate cancer. Unlike EGF, the IGF pathway may be equally important in the development of early prostate cancer. The IGF signaling pathway is a complex balance of interactions between IGF-1 ligand, multiple IGF binding proteins (IGFBPs), IGF receptor (IGFR), and IGFBP proteases. IGF-1 is synthesized in nearly every human tissue, but, in the prostate it appears to exert its action on prostate cancer cells through paracrine release from the prostate stroma (Moschos and Mantzoros, 2002). There are six known IGFBPs described in humans, and these compounds determine both the bioavailability of IGF-1 as well as guide its effects on target tissue (Djavan *et al.*, 2001; Moschos and Mantzoros, 2002). Of the circulating IGF-1, 99% is bound to IGFBPs with 75% of IGF-1 bound specifically to IGFBP 3. IGFR is also constitutively expressed in human tissues but quantitative receptor expression can be altered and can affect tissue response to IGF-1 (Moschos and Mantzoros, 2002; Krueckl *et al.*, 2004). IGF-1 function can be further regulated by IGFBP proteases, including PSA. In prostate cancer cells, binding of IGF-1 to IGFR initiates two predominant apoptotic resistance pathways: the PI3K/AKT pathway and, to a lesser extent, the NF-κB pathway (Moschos and Mantzoros, 2002; Bogdanos *et al.*, 2003). In a pattern similar to the EGF-EGFR signaling events, in the presence of elevated IGFR (common in advanced disease), IGF-1 can rescue prostate cancer cells from apoptosis induced by PI3K/AKT disruption (Miyake *et al.*, 2000). IGF-1 can directly stimulate the AR thus contributing to the emergence of androgen independence (Krueckl *et al.*, 2004). In addition, the differential expression of IGFBPs by prostate cancer cells impacts their sensitivity to apoptosis. While several of the IGFBPs have been implicated in prostate cancer progression, IGFBP 3 appears to be the most influential player (Hong *et al.*, 2002; Li *et al.*, 2003b), via its binding to IGF-1, attenuating the PI3K/AKT pathway and leading to increased apoptosis. Additional evidence shows that IGFBP 3 is able to sensitize prostate cancer cells to apoptosis in the absence of IGF-1 binding (Hong *et al.*, 2002). In view of the above evidence, it is not surprising that down-regulation of IGFBP 3 is commonly encountered in prostate cancer (Chan *et al.*, 1998a; Djavan *et al.*, 2001).

The contribution of the IGF/IGFB/IGFR/PSA axis in prostate cancer progression to metastatic disease is tremendous. Increased IGFR expression is common in androgen-independent metastatic lesions, and increased IGFBP 2 and 5 in human prostate tumors correlate with increased Gleason grade (Djavan *et al.*, 2001). Furthermore, elevated serum IGF-1 levels, as well as an elevated IGF/IGFBP 3 ratio have been identified by multiple clinical studies to be independent predictors of prostate cancer risk and to improve the diagnostic yield of PSA screening (Li *et al.*, 2003b; Stattin *et al.*, 2004). In the context of the bone microenvironment, the IGF-1 axis has been identified as the predominant survival factor pathway responsible for the androgen ablation and chemotherapy failure seen in prostate cancer bony metastasis (Bogdanos *et al.*, 2003). As metastatic prostate cancer cells release urokinase-type plasminogen activator (uPA), hydrolysis of IGFBPs occurs, resulting in a local increase in IGF-1 bioavailability and subsequent apoptotic resistance and osteoblastic reaction (Bodganos *et al.*, 2003; Miyamoto *et al.*, 2004). Similar treatments aimed at disrupting IGF signal transduction including GNRH antagonism, somatostatin analogs, and IGFBP protease inhibition, are currently under active investigation. To date, while quality of life measurements with such therapeutic strategies are encouraging, no significant changes in survival have been appreciated (Bodganos *et al.*, 2003).

The role of transforming growth factor-β1 (TGF-β1) in prostate cancer pathogenesis is reflected in a classic signal transduction mechanism, dominated by an abundant apoptosis-inducing ligand towards cell death evasion linked to disease progression. TGF-β1 signaling in normal prostate epithelium and in early prostate cancer leads to apoptosis induction and tumor suppression (Jones, Pu and Kyprianou, 2008). The ability of TGF-β1 to suppress early prostate cancer tumorigenesis requires intact signal transduction via interaction with the TGF-β1 receptors transforming growth factor β receptor-I (TGFβR-I) and transforming growth factor β receptor-II (TGFβR-II), and subsequent downstream targeting through regulation of the Smad family of protein effectors (Bello-DeCampo and Tindall, 2003). Up-regulation of this pathway from TGF-β1 and receptor binding leads to caspase-1 activation, up-regulation of BAX and down-regulation of BCL-2, ultimately resulting in tumor cell apoptosis (Guo and Kyprianou, 1999; Bruckheimer and Kyprianou, 2002). Furthermore, detection of enhanced expression of the ligand, TGF-β1, and its receptors after medical or surgical castration has been implicated as the main driving force for the pronounced prostate cancer cell apoptosis seen with such therapy (Wikstrom *et al.*, 1999).

Increased TGF-β1 ligand expression directly correlates with prostate cancer progression, while there is loss of expression in its receptors (Guo and Kyprianou, 1998; Wikstrom *et al.*, 1999). Disruption of normal TGF-β1 signaling tips the axis in favor of enhanced angiogenesis, extracellular matrix remodeling favorable for invasion and, most importantly, immunosuppression (Matthews *et al.*, 2000; Tuxhorn *et al.*, 2002c). TGF-β1 overexpression, common in advanced prostate cancer, directly inhibits the ability of tumor-specific cytotoxic T lymphocytes (CTLs) and NK cells to induce prostate cancer cell apoptosis; down-regulation of TGF-β1 restores immunogenicity of prostate cancer cells and suppresses metastasis formation (Matthews *et al.*, 2000; Shah *et al.*, 2002), potentially via induction of IL-6, a powerful inhibitor of prostate cancer cell apoptosis and metastasis promoter (Park *et al.*, 2003). BCL-2 overexpression can also inhibit TGF-β1 induced apoptosis (Bruckheimer and Kyprianou, 2002). Moreover, TGF-β1 can synergize with AR transactivation in response to androgen and up-regulate downstream targets, such as PSA, which have been implicated in apoptotic evasion (Kang *et al.*, 2001).

TGF-β1 overexpression coupled with the down-regulation of TGFβR-I and TGFβR-II are hallmarks of advanced prostate cancer. Numerous clinical studies using prostatectomy specimens and serum analysis of patients both before and after prostatectomy have revealed that up-regulation of TGF-β1 along with down-regulation of TGFβR-I and TGFβR-II is associated with invasive disease, increased Gleason grade and castration-resistant prostate cancer (Shariat *et al.*, 2004a; Zeng *et al.*, 2004). The prognostic power of TGF-β1 is exemplified by its inclusion in current pre-operative nomograms that have proven more effective at predicting recurrent disease than standard parameters used today, such as pre-operative PSA or Gleason grade (Kattan *et al.*, 2003). Selective approaches to target TGF-$\beta$1 signaling include quinazoline-based $\alpha_1$-adrenoceptor antagonists, restoration of TGF-$\beta$ receptor expression through gene delivery, and antisense inhibition of TGF-$\beta$1 expression (Matthews *et al.*, 2000; Partin, Anglin and Kyprianou, 2003; Jones, Pu and Kyprianou, 2008).

### 3.11.3 *Inflammation*

Acute inflammation is a rapid and self-limiting process as chemical mediators are induced in a tightly regulated sequence, and immune cells move in and out of the affected area destroying infectious agents, repairing damaged tissue, and initiating a specific and long-term response. Chronic inflammation reactions have been linked to the pathophysiology of human cancer, including prostate cancer (De Marzo *et al.*, 1999). Indeed, emerging insights into the molecular pathogenesis of prostate cancer suggest that damage to the prostate epithelium, potentially inflicted by diverse environmental exposures (such as infections and trauma), and/or microenvironment stimuli (such as oxidative stress and hypoxia), triggers pre-malignant inflammatory processes to promote tumor development and progression. In this milieu, the damaged epithelium may generate proliferative inflammatory atrophy (PIA) lesions, which may progress through prostatic intraepithelial neoplasia (PIN) to prostate cancer. Tumor stroma comprises fibroblasts, vascular, and lymphatic endothelial cells and smooth muscle cells surrounding the blood vessels. Significant populations of infiltrating leukocytes are present in both the supporting stroma and among the tumor cells, with macrophages and T cells differentially distributed among the tumor mass (Coussens and Werb, 2002). Tumor-associated macrophages (TAMS) are key cells in chronic inflammation and form a large part of the tumor mass in many human malignancies (Condeelis and Pollard, 2006). Initially recruited by the inflammatory cytokines, TAMs are provided with growth factors and survival signals by the tumor microenvironment. It is a very dynamic chemical conversation that leads to a highly aggressive malignant phenotype. Our current understanding of the contribution of inflammation to the tumorigenic process points to an enticing question: is it possible that the immune response generated to combat cancer initiation and progression provides yet another opportunistic interaction within the tumor microenviron-ment? While a connection between inflammation, and cancer development and progression has been suspected for some time, recent evidence has provided new insights into the important dynamic of this relationship. Two experimental models have identified the inflammatory response, occurring both in intes-tinal colitis and chronic hepatitis, as a key mediator in the development and progression of solid tumors (Balkwill and Coussens, 2004). In both a colitis and hepatitis model, the NF-κB system appears to be the intracellular signaling link that allows the inflammatory response to be a potential co-conspirator in tumor progression (Viatour *et al.*, 2005). Further dissection by Greten and colleagues identified the secretion inflammatory mediators such as TNF-α, interleukin-1 (IL-1), interleukin-6 (IL-6) and interleukin-8 (IL-8) driving the NF-κB pathway towards apoptotic resistance and tumor progression (Gretten *et al.*, 2004). The mechanisms by which *Helicobacter pylori*-induced chronic inflammation causes gastric cancer include oxidative stress, perturbation of the growth balance between cell proliferation, apoptosis among the epithelial cell populations, and secretion of inflammatory cytokines (Bjornsti and Houghton, 2004b). Injury to the prostate epithelium, regardless of the source of damage, elicits a stereotypical stress and regenerative response characterized by the emergence of PIA lesions as described by De Marzo and colleagues (De Marzo *et al.*, 2007). Considering the evidence linking prostate cancer development to chronic inflammation (Konig *et al.*, 2004; Nelson *et al.*, 2004), one recognizes the need for molecular exploration of this relationship. Compelling molecular evidence suggests a mechanistic association between inflammation and obesity genes in prostate cancer (Wang *et al.*, 2009), as well as the appearance of somatic epigenetic alterations in PIA lesions that closely resemble those present in prostate tumors (Bardia *et al.*, 2009). The most prominent one is the hyper-methylation of cytosine bases in the regulatory region of the gene encoding the π–class glutathionine S-transferase (GSTP1), resulting in transcriptional silencing of the gene and enhanced susceptibility to various tumor-inducing genomic stress signals (Yegnasubramanian *et al.*, 2008).

In the clinical setting, analysis of two major prostate tumor-infiltrating lymphocyte subsets revealed that capsular and perineural invasion, as well as biochemical progression, were related to strong expression of T and B cells (Karin, 2006). Specifically, a high density of CD4+T helper-inducer lympho-cytes located in areas of prostate adenocarcinoma has been associated with poor survival and higher Gleason scores (McArdle *et al.*, 2004). In addition, there is an inverse relationship between the degree of primary tumor infiltration by CD68 macrophages and the likelihood of prostate cancer progression (Shimura *et al.*, 2000).

Perhaps the most important lesson learned so far from the experimental and clinical studies on prostate cancer is the vital role of cytokines in the host inflammatory response during tumor progression (Karin, 2006). There are two cytokines most often implicated in this dual capacity, TNF-α and IL-6. Tumor necrosis factor-α (TNF-α) is a pleiomorphic cytokine heavily involved in inflammation. The cellular response to TNF-α ligand-receptor binding can invoke either the apoptotic cascade or promote tumor cell survival. There are two well-described TNF-α receptors, TNFRI and TNFRII. Ligand binding to TNFRI usually results in FADD recruitment and subsequent caspase-8 activation, which ultimately results in apoptosis (Guseva *et al.*, 2004). Ligand binding to TNFRII leads to activation in the MAPK and NF-κB pathways resulting in survival and apoptotic resistance. However, receptor expression alone does not dictate the fate of the tumor cell, as TNFRI binding can also stimulate NF-κB activation through TRAF2 activation of IKK-mediated IκB-α phosphorylation (Chopra *et al.*, 2004; Guseva *et al.*, 2004). Prostate cancer cellular response to TNF-α appears to be linked to androgen responsiveness (Mizokami *et al.*, 2000). Indeed, experimental evidence confirms that androgen-responsive prostate cancer cells are sensitive to TNF-α-induced apoptosis via both p53 accumulation as well as BID cleavage and subsequent caspase cascade initiation leading to cytochrome c release (Rohklin *et al.*, 2000; Kulik, Carson and Vomastek, 2001). In androgen-independent prostate cancer cells however, TNF-α promotes apoptotic resistance rather than sensitivity. This axial shift towards tumor promotion (from apoptosis) is attributed to the high constitutive NF-κB activation in androgen-independent prostate cancer cells, as well as TNF-α-mediated up-regulation of IKK activity and subsequent NF-κB activation through PI3K/ AKT-dependent and independent pathways (Muenchen *et al.*, 2000; Gustin *et al.*, 2001; Dhanalakshmi *et al.*, 2002; Chopra *et al.*, 2004).

Increased TNF-α expression has been reported in both epithelial tumor cells and tumor-associated macrophages (Michalaki *et al.*, 2004). Furthermore, serum levels of TNF-α from prostate cancer patients, correlated with both bulky local disease and metastatic progression, potentially serve as an independent prognostic indicator for survival (Michalaki *et al.*, 2004). Due to the pleomorphic response of prostate cancer cell lines to TNF-α, this intriguing molecule acquired wide therapeutic value. Therapeutically, the radiation-sensitizing effects of TNF-α have been interrogated in multiple tumor systems including prostate cancer (Chung *et al.*, 1998; Kimura *et al.*, 1999). Mechanistically, targeted inhibition of the NF-κB pathway has been shown to drive the response to TNF-α downstream in the apoptotic pathway (Papandreou and Logothetis, 2004). An emerging therapeutic weapon in the cancer armamentarium, proteosome inhibition has been implicated in targeting NF-κB activation, with promising results in patients with castration-resistant prostate cancer (Papandreou and Logothetis, 2004).

Similarly, IL-6, while traditionally described as a key mediator in the inflammatory response, has proven to be an integral part of cancer responses to apoptosis stimuli. Elevations in IL-6 can activate the PI3K/AKT pathway with subsequent up-regulation of BCL-XL (Mutsaers *et al.*, 1997; Pu *et al.*, 2004; Xie *et al.*, 2004), and resistance to standard chemotherapy-induced apoptosis. This enhanced AKT activation has also been associated with neuroendocrine differentiation, a common phenotype of androgen-independent prostate cancer cells (Xie *et al.*, 2004). IL-6 has also been shown to stimulate AR activity in the absence of androgen via the stat3 signaling pathways. This ability to bypass the receptor ligand interaction of androgen and its receptor allows IL-6 to protect androgen-sensitive prostate cancer cells from apoptosis induced by androgen deprivation (Lee *et al.*, 2004). Interestingly, there is an inversely proportional correlation between IL-6 levels and androgen expression in aging, healthy male subjects, and the effects of IL-6 on hormone responsive cell lines can be blunted with the addition of androgen (Kim *et al.*, 2004; Xie *et al.*, 2004). IL-6 overexpression in the prostate tumor microenvironment is due to both autocrine and paracrine feedback loops. The constitutive overexpression of IL-6 in androgen-independent prostate cancer cells is attributed to an autocrine loop governed by the NF-κB activity (Zerbini *et al.*, 2003). In metastasis to the bone, the microenvironment engages IL-6 to facilitate the paracrine interactions of osteoblasts with prostate tumor cells (Garcia-Moreno *et al.*, 2002).

To attenuate prostatic tumorigenesis driven by chronic or recurrent prostate inflammation, rational chemoprevention has thus far featured anti-inflammatory drugs and antioxidants. Aspirin and non-aspirin non-steroidal anti-inflammatory drugs (NSAIDs), which act to reduce inflammation by inhibiting cyclooxygenases, have received wide attention as potential chemopreventive agents for many cancers including prostate cancer (Cuzick *et al.*, 2009). The clinical significance of IL-6 is exemplified in its prognostic capabilities. Studies including serum measurements of IL-6 with or without the addition of its soluble receptor have shown both to be powerful predictors of PSA failure, disease progression, and mortality (Shariat *et al.*, 2001; George *et al.*, 2005). Furthermore, pre-operative serum IL-6 measurements appear to be an effective screening tool for occult metastatic disease at the time of resection (Shariat *et al.*, 2004b) and may prove valuable in the development of adjuvant therapy protocols. Attempts to directly target IL-6 expression with monoclonal antibody therapy have had moderate success in animal models (Smith and Keller, 2001). One could easily argue that as autocrine IL-6 production is apparently NF-κB-driven, NF-κB targeting and proteosome inhibition may ultimately inhibit this cytokine's pro-survival activity as well.

# 4

# Androgen Receptor-Mediated Apoptosis: Significance in Development of Castration-Resistant Prostate Cancer

## 4.1 The Androgen Receptor (AR)

Androgen exerts its biological functions through an axis involving testicular synthesis of testosterone, conversion by $5\alpha$-reductase to the active metabolite $5\alpha$-dihydrotestosterone (DHT) and its binding to the androgen receptor (AR) to induce transcriptional activation of target genes (Siiteri and Wilson, 1974; Mainwaring, 1977; Imperato-McGinley et al. 1985; Heinlein and Chang, 2002). The AR is a 110kDa phosphoprotein and a member of the nuclear receptor super-family. It shares a common tetramodular structure with other members of the nuclear receptor family, consisting of an N-terminal transactivation domain, a central DNA binding domain, a hinge region, and a C-terminal ligand-binding domain. In the cytoplasm, AR is complexed with molecular chaperones from the heat shock protein family hsp70, hsp90, and hsp56, which maintain the AR in a conformation conducive to ligand binding (Ueda et al., 2002; Heinlein and Chang, 2002). The liganded AR translocates to the nucleus and binds, along with co-regulators, to the androgen response elements (AREs) in the promoter/enhancer region of target genes, leading to transcription and downstream effects (Yeh et al., 1999; Centenera et al., 2008). Once bound to DNA, AR recruits transcriptional co-activators to assist with gene expression (Agoulnik and Weigel, 2008). The contribution of AR to prostate cancer initiation and progression is incontrovertible. As discussed in the introductory chapter, the epithelial compartment of the human prostate is composed of two types of cells

— basal epithelial cells, and glandular epithelial secretory cells which comprise the predominant cell type. In the adult prostate, androgens promote survival of the epithelial cells, the primary step to malignant transformation to prostate adenocarcinoma (De Marzo *et al.*, 1998). Prostate tumors are initially androgen responsive, hence the standard treatment approaches for metastatic disease uses combined androgen blockade, androgen deprivation, and direct antagonism of the AR with anti-androgens (Grossman, Huang and Tindall, 2001).

In recent years there has been substantial progress in elucidating the molecular details of castration-resistant prostate cancer (CRPC), with near consensus in the field that CRPC develops as a result of restored AR signaling (Edwards *et al.*, 2003). Castration-resistant prostate cancer is commonly associated with increased AR gene expression, which occurs through AR gene amplification or mutational activation (Knudsen and Scher, 2009). Is elevated AR content necessary and sufficient to confer resistance to anti-androgen therapy? Indeed, explicit characterization of castration-resistant prostate cancer at the molecular level and in model systems validated the concept that AR activity is regained during tumor progression and metastatic disease manifestation (Gregory *et al.*, 2001a). Once disease progression is documented, the course is inevitably fatal. But to what extent does the absence or mutational status of the AR contribute to the therapeutic failure of prostate cancer patients with advanced disease? There is a popularized notion that castration-induced apoptosis-based treatment of advanced prostate cancer is limited in its therapeutic efficacy, and impact on patient survival, by the development of resistance among the epithelial cell populations to androgen deprivation-induced apoptosis due to loss of AR function, and this has been debated in recent years. Androgen depletion therapies and direct AR antagonists rely on an intact AR ligand-binding domain and stand on the premise that agonist binding to the ligand binding domain is universally required for AR activation. The emerging scenario initiated by studies reported by Don Tindall's group demonstrated nicely that AR can be alternatively spliced so that the C-terminal domain is deleted, rendering production of a receptor that is constitutively active (Dehm *et al.*, 2008). Consistent with these studies in experimental models, constitutively active splice variants (lacking the C-terminus) have been reported in castration-resistant prostate cancer specimens (Hu *et al.*, 2009). These exciting new findings point to the need for the development of a new class of AR-inhibitory agents for the effective treatment of prostate tumors expressing truncated AR, where the classic, complete androgen ablation would have no impact on the AR activity.

## 4.2 Androgen Ablation: The Glory and the Failures

The striking dependence of prostatic adenocarcinoma development on AR signaling, at all stages of disease, has been heavily exploited in the treatment of prostate cancer for the last three decades after being pioneered by Dr Charles Huggins (Huggins, 1967). Non-steroidal anti-androgens (flutamide, bicluta-mide, and nilutamide) are competitive inhibitors for the AR. Since they do not produce the serum testosterone levels, they do not produce all the toxicity and side effects associated with castration (erectile dysfunction, loss of libido, muscle wasting, and osteoporosis). While initially the vast majority of meta-static prostate tumors rely on the availability of androgens for growth and survival, in the final stages of the disease, patients with metastatic prostate cancer eventually progress clinically under androgen-deprived conditions (Tso *et al.*, 2000). Hence the name officially describing advanced prostate cancer as castration-resistant prostate cancer (CRPC); that is, resistant to the two primary treatments of metastatic prostate cancer: castration-induced androgen depletion using androgen lowering agents such as leuprolide (Lupron), and AR antagonists such as bicalutamide (Casodex) (Eisenberger *et al.*, 1998). Direct AR antagonists are frequently clinically utilized in combination with castration or LH-RH analogs in an effort to inhibit AR signaling. Although these strate-gies are initially effective, recurrent tumors arise with restored AR activity, and no durable treatment exists to combat advanced disease. Docking of AR antagonists such as bicalutamide into the AR C-terminal ligand-binding domain (LBD) results both in passive AR inhibition (via competition for agonists), and an active mechanism of AR inhibition (prevention of co-activator binding and/or co-repressor recruitment) (Shang, Myers and Brown, 2002). Novel insights into AR regulation and the mechanisms underlying resurgent AR activity provide a most fertile forum for the development of novel strategies to more effectively inhibit AR activity and prolong the transition to therapeutic failure (Wang *et al.*, 2007). Elegantly designed pre-clinical studies by Charles Sawyer's group demonstrated that elevated AR expression was necessary and sufficient for CRPC growth in xenograft models, pointing to the critically important point that CRPC remains AR-dependent (Chen *et al.*, 2004). Striking evidence is rapidly accumulating to indicate that AR activity is restored in CRPC by various mechanisms including AR gene amplification, AR gene mutation, increased intratumoral testosterone synthesis (Montgomery *et al.*, 2008), and cross-talk with various kinase signaling pathways (Craft *et al.*, 1999).

Therapeutic failure to androgen ablation therapy is heralded first by increasing PSA levels and recurrent prostate tumor growth (Pienta and Bradley, 2006), followed by metastatic spread to the bone, and advanced disease symptoms. In the pre-PSA era, approximately 40–50% of men with prostate cancer had bone metastases at first clinical presentation. Bone metastases arising from prostate cancer are a major unresolved clinical problem, and despite the best efforts of clinicians and basic researchers, the outcome of treatment has little changed since the first description of this condition in the modern medical literature (Brodie, 1834; Scott, 1953). Morbidity and death from prostate cancer are usually the direct result of invasion of the bone marrow by malignant cells, and the consequent disturbance of the integrated structure and function of the bone marrow and the skeletal bones (Shahinian *et al.*, 2005). The sequence of events leading to bone metastases from a primary prostate tumor is initiated by cellular binding to, and subsequent invasion of, the lining endothelium of red blood vessels by circulating tumor prostate cancer epithelial cells (Jacobs, 1983).

## 4.3 AR Status in Castration-Resistant Prostate Cancer

The absence of a link between elevated serum testosterone, DHT, or adrenal androgens, and prostate cancer risk suggests that androgens are not sufficient to promote prostate carcinogenesis (Roberts and Essenhigh, 1986; Hsing, 2001). That being stated, could the receptor (AR) bypass the requirement of the ligand? Studies evaluating androgen levels in prostate tissue showed that androgen depletion therapies reduced intratumoral androgens by only 75% at a time when serum androgen levels remained in the castrate range (Montgomery *et al.*, 2008). Such a persistent expression of AR and AR target genes, for example, PSA and TMPRSS2, in prostate tumor specimens, supports the concept that the residual androgen is sufficient to sustain AR activity. Moreover, circulating tumor cells isolated from patients with CRPC have evidence of AR amplification in 50% of cases, further validating a role for AR as a major contributor to the emergence of CRPC (Vishnu and Tan, 2010). Abberant AR gene activation can then cleverly enhance the ability of the receptor to "adapt" to the environment of low androgen levels, in part by using low-affinity ligands (Chen *et al.*, 2004). Aberrant activation of AR can result from gene amplification, mutation, phosphorylation, activation of co-regulators, or androgen-independent activation. Most cases of prostatic carcinoma resistant to androgen ablation therapy demonstrate activation of AR by one of the following mechanisms (as reviewed by Knudsen and Scher, 2009). The AR gene is either mutated or amplified in 20–30% of androgen-resistant prostate carcinomas. Further, 20% of castration-resistant carcinomas contain gene amplifications as compared to just 2% of hormone-sensitive tumors, suggesting that aberrant activation in response to low levels of androgens or other ligands may underlie the progression to aggressive disease that is refractory to androgen ablation therapy. Specific mutations, such as the T877A and H874Y substitutions, confer increased sensitivity to AR for steroid hormones such as progesterone, 17β-estradiol, or hydroxyflutamide in prostate cancer cell lines and xenografts. Mutant AR containing the E231G substitution predisposes transgenic mice to the development of prostatic intraepithelial neoplasia (PIN), adenocarcinoma, and metastases. AR hyperactivity leads to prostate cancer initiation and progression to advanced disease: overexpression of AR leads to the development of focal PIN, whereas AR overexpression in LAPC-4 prostate cancer cells and xenografts results in a transition from androgen-sensitive disease to androgen-resistant cancer (Chen *et al.*, 2004).

The AR has distinct functional domains that include the LBD, a DNA-binding domain (DBD), and an amino terminal domain (NTD). AR NTD contains the AF-1 that contributes most of the activity to the ligand-bound AR (Jenster *et al.*, 1991; Simental *et al.*, 1991), rather than AF-2 in the LBD, making the AR unique from other steroid receptors. The AR NTD is activated by alternative pathways in the absence of androgen (Sadar, 1999; Ueda, Bruchovsky and Sadar, 2002; Quayle, *et al.*, 2007). Binding of androgens to the LBD of AR results in receptor activation, so it can effectively bind to its specific DNA consensus site (the androgen response element (ARE)) on the promoter and enhancer regions of androgen regulated genes, towards transcription initiation. Evidence supporting the AR in CRPC includes the fact that many of the same genes activated in response to androgens in androgen-dependent prostate cancer become elevated in CRPC (Gregory *et al.*, 1998). This suggests AR activation in the absence of testicular androgens in CRPC, consistent with the notion of nuclear AR presence in metastatic prostate cancer tumors (Gregory *et al.*, 2001b; Kim *et al.*, 2002). There is also evidence to suggest delayed onset of CRPC by altering the timing and sequence of use of anti-androgens (Bruchovsky *et al.*, 2001), and the necessity of AR for proliferation and tumor growth (Chen *et al.*, 2004). The mechanism of AR transactivation in the absence of androgens by alternative pathways such as cAMP/PKA, IL-6, and epidermal growth factor, involves the AR NTD (Sadar, 1999; Ueda, Bruchovsky and Sadar, 2002; Gregory *et al.*, 2004; Quayle *et al.*, 2007), and CRPC is independent of the AF-2 region in the LBD (Dehm and Tindall, 2006). Thus, in the absence of androgens, the AR NTD can be activated and can contribute mechanistically to CRPC. The support for this functional contribution of AR to CRPC is built on compelling evidence from different investigators describing naturally occurring splice variants of the AR that lack the LBD in prostate cancer cell lines (LNCaP and 22Rv1), and in CRPC (Dehm *et al.*, 2008; Guo *et al.*, 2009; Hu *et al.*, 2009). These mutants are constitutively active, and resist inhibition by current therapeutic strategies targeting the AR LBD, such as anti-androgens and androgen ablation therapy. The clinically available anti-androgens targeting the LBD predominantly fail, presumably due to poor affinity and gain-of-function mutations in the LBD that lead to activation of the AR by these same anti-androgens (Taplin *et al.*, 1999).

The AR NTD is essential for AR activity, a role justifying its functional involvement in the underlying molecular mechanism of CRPC. In the absence of androgen, full-length AR is activated through its NTD by factors secreted from bone and by alternative pathways. The elegant studies recently reported by Marianne Sadar's group identified an antagonist to the AR NTD that interacts with the NTD and impairs protein–protein interactions with the NTD, causing a marked suppression of CRPC tumor via apoptosis induction (Anderson *et al.*, 2010). Changes in protein–protein interactions of the AR NTD by stimulation of alternative pathways may mediate a transcriptionally active receptor in the absence of ligand. Precedence for the concept that the agonist ligand is not essential for receptor activation, providing the molecular basis for small molecule regulation, is derived from other steroid receptors such as the estrogen receptor (Wang *et al.*, 2006). The AR NTD is flexible with a high degree of intrinsic disorder with characteristics of a collapsed disordered conformation (Lavery and McEwan, 2008). The recent identification of the small molecule inhibitor EPI-001, that inhibits both ligand-independent and ligand-dependent interaction with CBP in a highly specific manner, generated great momentum for prostate cancer therapeutics in the pharmacological drug-discovery arena. The selectivity of EPI-001 to inhibit transactivation of the AR NTD and attenuate AR activity through blocking protein–protein interactions and reducing binding to AREs has significant therapeutic potential for the treatment of patients with CRPC (Anderson *et al.*, 2010). There is "light at the end of the tunnel" in the pursuit of AR therapeutic targeting in CRPC.

## 4.4  AR Interactions with Growth Factor Signaling Leads to Apoptosis

Extranuclear pools of steroid receptors exist in cytoplasmic organelles and localize to plasma membranes, primarily in caveola rafts (Hammes and Levin, 2007). The localization of all sex steroid receptors to the plasma membrane rafts is controlled in a very dynamic manner via the process of palmitoylation in a highly conserved sequence (Pedram *et al.*, 2007). At the plasma membrane, the AR (like the other steroid receptors) physically interacts with multiple signaling molecules, scaffold and linker proteins, and effects the activation of discrete G protein $\alpha$ and $\beta\gamma$ subunits. G protein activation results in the transduction of cyclic nucleotides, calcium and kinase cascades, occurring as very early response events, that is, seconds to minutes after receptor binding by the ligand. Such receptor-membrane-initiated signals cause the phosphorylation of substrate proteins to alter their cellular localization and functional activity. Membrane AR is well established to mediate rapid signaling through ERK, Src kinase and PI3K towards cell proliferation and survival (Migliaccio *et al.*, 2000; Freeman *et al.*, 2007). Recently published functional studies demonstrate that the heat shock protein 27 (Hsp27) promotes the palmitolyation and subcellular localization of endogenous AR at the membranes of prostate cancer cells LNCaP (Hsp27) (Razani *et al.*, 2010). This evidence implicates Hsp27 as a modulator of survival and apoptosis signaling via engaging AR in androgen-sensitive prostate tumors. This is a very appealing rationale for orchestrating precise targeting strategies using membrane-specific AR agonists and antagonists to trigger apoptosis in both androgen-dependent and androgen-independent prostate cancer.

Androgen-induced prostate epithelial cell proliferation, besides the AR requirement, is regulated by an indirect network of signaling pathways, spatially coordinated be the tissue microenvironment, and involving paracrine mediators produced by stromal cells, such as insulin-like growth factor (IGF), fibroblast growth factor (FGF), and epidermal growth factor (EGF) (Cunha and Donjaour, 1989; Byrne, Leung and Neal, 1996; Zhang, Magit and Sager, 1997). The literature is overflowing with evidence on the dysfunctional regulation at the transcriptional or translational level of these signaling effectors as critical contributors to ligand-independent AR activity. The major studies connecting aberrant activation of AR with the key growth factor signaling pathways in the context of apoptosis regulation during prostate tumorigenesis are discussed here.

### 4.4.1 *AR connects with EGF*

EGF and its membrane receptor, the EGFR, are involved in the pathogenesis of different tumors, including prostate cancer (Russell, Bennett and Stricken, 1998). Both the ligand and its signaling receptor partner are frequently up-regulated in advanced stages of prostate cancer (Di Lorenzo *et al.*, 2002). Targeting EGFR with monoclonal antibodies, or with tyrosine kinase inhibitors, suppresses growth and invasion of androgen-dependent and -independent prostate cancer cells *in vitro* (Bonaccorsi *et al.*, 2004b; Festuccia *et al.*, 2005). The involvement of EGFR in proliferation and invasion of cancer cells has been strongly supported by growing evidence (Wells *et al.*, 2002). EGFR also participates in the formation of plasma membrane structures (lamellipodia), which mediate migration through the basal membrane (Rabinovitz, Gipson and Mercurio, 2001). Significantly, elevated EGFR enhances the invasion potential of mammary tumors, by increasing cell motility without affecting tumor growth (Xue *et al.*, 2006), emphasizing the critical role exerted by the EGF/EGFR system in invasion and metastasis. Moreover, the robust evidence on the interaction between EGF/EGFR and androgen signaling provides proof-of-principle that engagement of multi-crossed signals is crucial for the acquisition and maintenance of androgen sensitivity (Oosterhoff, Grootegoed and Blok, 2005; Léotoing *et al.*, 2007). Expression of the androgen-regulated PSA gene is induced by the administration of IL-6, which activates EGFR (Hobisch *et al.*, 1998; Ueda *et al.*, 2002). Furthermore, androgen-dependent prostate cancer cell proliferation is impaired by exposure to an EGF receptor kinase inhibitor (Vicentini *et al.*, 2003). This evidence initially pointed to the contribution of EGFR in dictating AR outcomes in prostate cancer cells. ErbB2, a lead member of the epidermal growth factor receptor family of receptor tyrosine kinases, was shown to be overexpressed in prostate cancer during progression to androgen-independent metastatic disease (Heinlein and Chang, 2004). The mechanistic basis for the AR and ErbB2 cross-talk is provided by other reports, indicating that modulation of AR signaling activity by the HER-2/new tyrosine kinase promotes androgen-independent prostate tumor growth *in vitro* and *in vivo* (Craft *et al.*, 1999; Yeh *et al.*, 1999). Loss-of-function studies strongly support such a signaling interaction by indicating the modulatory function of HER2/new kinase on AR activity through an impact on DNA binding and stability, ultimately impairing prostate cancer cell growth (Mellinghoff *et al.*, 2004). Taken together, these lines of evidence converge in recognition of the ErbB2 kinase activity being required for optimal transcriptional activity of AR in prostate cancer cells (Mellinghoff *et al.*, 2004; Liu *et al.*, 2005a).

Androgens can control protein expression post-transcriptionally by regulating the binding of endogenous HuR to the AU-rich 3′UTRs of, for example, EGF mRNA (Myers *et al.*, 1999; Torring *et al.*, 2003). The ability of androgens to regulate the expression of ARE-binding proteins that bind to these instability elements supports an additional mechanistic involvement (by androgens) in the post-transcriptional control of EGF (Simon and Toomre, 2000; DiNitto, Cronin and Lambright, 2003; Kuhajda, 2006). In a "reversal-of-action" mode, EGF reduces AR expression and blocks androgen-dependent transcription in differentiated cells, while it activates the AR promoter (Culig *et al.*, 1994). This intriguing EGF-AR interplay is an important contributor to prostate tumor progression, but it is not exclusive to EGF, as AR activity can be modulated by other growth factors (Orio *et al.*, 2002).

AR interacts with the MAPK/extracellular signaling-regulated kinase-1 (MEKK1) and the epidermal growth factor-1 receptor (EGFR) (Abreu-Martin *et al.*, 1999; Bonaccorsi *et al.*, 2004a) resulting in apoptosis activated mitogen-activated protein kinase (MAPK) (Peterziel *et al.*, 1999) and in a "functional-symmetry", EGF-activated MAPK signaling cascade interferes with AR function, modulating the androgen response. Mitogen and extracellular kinase (MEK) inhibition reverses the EGF-mediated AR down-regulation in differentiated cells, thus suggesting the existence of an inverse correlation between EGF and androgen signaling in non-tumor epithelial cells (Léotoing *et al.*, 2007). Additional key signal transducers in this dynamic include transducer activator of transcription 3 (Stat3), most probably required for AR activation by IL-6 towards promoting metastatic progression of prostate cancer (Abdulghani *et al.*, 2008). Stat3 increases the EGF-induced transcriptional activation of AR, while androgen pre-treatment increases Stat3 levels in an IL-6 autocrine/paracrine dependent manner suggesting an intracellular feedback loop (Aaronson *et al.*, 2007). AR can also affect clathrin-mediated endocytosis pathway of EGFR, an essential step in its signaling integrity. The significance of engaging such a robust cross-signaling by prostate cancer cells towards determining their survival and response to the microenvironment is established by growing evidence (Bonaccorsi *et al.*, 2007).

The recently identified active integration of AR and EGFR signaling within the lipid raft microdomains in target cells provides an intriguingly topological twist to this cross-talk. Considering that the serine–threonine kinase AKT1 is a convergence point of the two hormonal stimuli, and that AR is localized in lipid raft membranes where it is stabilized by androgens (Freeman *et al.*, 2007), one could easily argue that the newly found membrane "domain" harboring AR is responsible for the non-genomic signaling by AR. The emerging concept that AKT1 is sensitive to manipulations in cholesterol levels gains direct support from biochemical analysis, verifying that a subpopulation of AKT1 molecules resides within lipid raft microdomains (Bauer *et al.*, 2003; Zhuang *et al.*, 2005). Distinct changes in the phosphorylation state of AKT1 in response to androgen occur quickly, but temporally independent in the raft and non-raft compartment, implicating processing of dissimilar signals. Interestingly, EGF triggers AKT1 phosphorylation via more rapid kinetics than those induced by androgens; this was recently documented by studies on the sensitivity of EGFR family proteins to disruptions in cholesterol synthesis and homeostasis, supporting the functional significance of EGF signal transduction through lipid rafts (Freeman *et al.*, 2007).

### 4.4.2 *AR and IGF interactions*

Signaling by insulin-like growth factor 1 (IGF-1) is of major mechanistic and biological significance (Burfeind *et al.*, 1996; Pollak, Beamer and Zhang, 1998; Wolk *et al.*, 1998; Nickerson *et al.*, 2001). The importance of insulin-like growth factor 1 (IGF-1) in prostate apoptosis has been supported by diverse lines of evidence in experimental models. Nickerson and colleagues first demonstrated *in vivo* that prostate cancer progression to androgen independence (castration-resistant disease) is associated with an up-regulation of IGF-1 and IGF-1 receptor (IGF-1R) (Nickerson *et al.*, 2001). Directly resonating with this concept, more recent evidence suggested that castration-induced androgen depletion apoptosis of human prostate tumor epithelial cells correlated with reduced IGF-1 levels (Ohlson *et al.*, 2007). In a scenario fostering AR reactivation in a low androgen environment (Grossmann, Huang and Tindall, 2001), insulin resistance, and hyperinsulinemia correlate with an increased incidence of prostate cancer (Fan *et al.*, 2007). High IGF-1 levels in the serum correlate with an increased risk of prostate cancer (Pollak, Beamer and Zhang, 1998; Wolk *et al.*, 1998), whereas IGF-1 enhances AR transactivation under low/absent androgen levels (Culig *et al.*, 1994; Orio *et al.*, 2002) and promotes prostate tumor cell proliferation (Burfeind *et al.*, 1996).

Endogenous AR expression, as well as AR transcriptional activity, is regulated by insulin via activation of the phosphatidylinositol 3-kinase (PI3K) transduction pathway (Manin *et al.*, 1992; Manin *et al.*, 2000; Manin *et al.*, 2002). As a downstream molecule, Foxo1 becomes phosphorylated and inactivated by PI3K/AKT kinase in response to IGF-1 or insulin, and subsequently suppresses ligand-mediated AR transactivation. Foxo1 is recruited by liganded AR to the AR promoters, and interacts directly with the C terminus of AR in a ligand-dependent manner, disrupting ligand-induced AR nuclear compartmentalization. This Foxo1 interference with AR-DNA interactions suppresses androgen-induced AR activity, resulting in prostate tumor suppression (Fan *et al.*, 2007).

An intracrine positive feedback between IGF-1 and AR signaling has been implicated in prostate cancer cells. Liganded AR up-regulates IGF-1 receptor expression in HepG2 and LNCaP cells, presumably resulting in higher IGF-1 signaling in prostate cancer cells. Two AREs within the IGF-1 upstream promoter activate IGF-1 expression (Wu *et al.*, 2007). In addition, androgens can control IGF signaling via modulation of IGF binding proteins (IGFBPs) in prostate epithelial cells, while both androgens and IGF-1 up-regulate IGFBP-5 mRNA in androgen-responsive human fibroblasts (Yoshizawa and Ogikubo, 2006). IGFBP-5 initially binds IGFs with high affinity, principally by an N-terminal motif, and inhibits IGF activity by preventing IGF interaction with the type 1 receptor (Kalus *et al.*, 1998). Taken together, this evidence supports a "higher-level" interaction between AR and the IGF signaling, via recruitment of direct pathways towards activation, transcriptional regulation, and protein post-translational changes, all critical to tumor cell survival.

### 4.4.3 *AR and TGF-β: partners in life and death (of the cell)*

Transforming growth factor-β (TGF-β) is a ubiquitous cytokine that plays a critical role in numerous pathways regulating cellular and tissue homeostasis (Shi and Massague, 2003). A member of a large family of multifunctional growth factors, TGF-β1 plays a major regulatory role in cell growth, apoptosis, and differentiation (Derynck and Zhang, 2003; Siegel and Massague, 2003). The active form of TGF-β is a 25-KD homodimer that is cleaved from a 110-KD inactive form; the activation of TGF-β is completed by proteases and results in a "sticky" molecule that can bind to various carriers (Massague and Gomis, 2006). TGF-β1 induces its biological effects through a family of transmembrane serine/threonine kinase receptors. Upon the ligand binding, TGF-β receptor II (TβRII) recruits TGF-β receptor I (TβRI), causing the phosphorylation of the TβRI GS domain and leading to the activation of TβRI kinase and L45 loop in the kinase domain for interaction with the Smad proteins. First, the Smad2/Smad3 proteins are phosphorylated, and upon interaction with Smad4 the complex promotes the nuclear translocation of the latter and induces transcriptional activation of target genes (Arsura *et al.*, 2003; Derynck and Zhang, 2003). Smad6 and Smad7 inhibit TGF-β signal transduction, by either promoting degradation of TGF-β receptors or acting as decoys in Smad-receptor or Smad-Smad interactions (Edlund *et al.*, 2005). Recent studies have identified multiple non-Smad pathways such as MAPK, LIMK, RhoA, and PI3K/AKT as functional intracellular effectors of the TGF-β signaling network (Moustakas and Heldin, 2005; Vardouli, Moustakas and Stournaras, 2005; Zhu *et al.*, 2006; Zhu *et al.*, 2010a). The dual nature of this multifunctional cytokine is intriguing: the Smad-dependent pathway might be acting as a tumor suppressor in early-stage prostate cancer, and then by changing into non-Smad pathways might act as a tumor promoter in advanced cancer (Akhurst and Derynck, 2001). In the normal prostate, or in the early development of prostate adenocarcinoma, the TGF-β ligand binds TGF-β receptor II and then the complex recruits TGF-β receptor I and subsequently leads to Smad protein activation to control target gene expression. A dysfunctional TGF-β signaling mechanism, due to expression loss/defects in transmembrane receptors and post-receptor effectors, results in loss of growth control thus contributing to prostate tumorigenesis (Shi and Massague, 2003; Bhowmick *et al.*, 2004). Primarily through paracrine effects (the stromal/tumor microenvironment), increased serum levels of TGF-β are found in patients with advanced prostate cancer; this TGF-β ligand increase is causally linked with increased angiogenesis, extracellular matrix degradation, epithelial-to-mesenchymal transition (EMT), and decreased immune surveillance, resulting in tumor invasion and metastasis (Akhurst and Derynck, 2001; Beck, Schreiber and Rowley, 2001; Bachman and Park, 2005; Jakowlew, 2006).

In normal prostate epithelial cells, TGF-β is a negative regulator of the growth dynamics due to its ability to inhibit proliferation and induce apoptosis, and also to regulate cell migration, adhesion, differentiation, and to mediate epithelial–stromal interactions in the context of the microenvironment (Tuxhorn *et al.*, 2002c; Edlund *et al.*, 2003). During the early stages of prostate tumor growth, TGF-β serves as a tumor suppressor by inducing cell cycle arrest and/or apoptosis. TGF-β has long been established as a physiological regulator of prostate growth because of its ability to induce apoptosis (Martikainen *et al.*, 1991; Kyprianou and Isaacs, 1989). The ability of TGF-β to cross-talk with other signaling pathways, mainly AR, enables a boost in the regulatory network of cell cycle progression and apoptosis (Zhu and Kyprianou, 2008). The androgen axis engages TGF-β in a dynamic cross-talk with the AR to regulate the myogenic differentiation via recruiting β-catenin and follistatin (Shah *et al.*, 2008). In prostate cancer cells, forced AR overexpression inhibits growth control function of TGF-β/Smad signaling (Zhu *et al.*, 2008). In the clinical setting, AR overexpression in hormone-refractory prostate tumors confers resistance to the growth-inhibitory effects of elevated serum TGF-β1, even in the absence of androgens (Van der Poel, 2008; Zhu and Kyprianou, 2008). Furthermore, TβRII is intimately involved in the androgenic control of TGF-β signaling in prostate epithelial cells (Song *et al.*, 2003). Pertaining to such function, the TGF-β promoter contains three distal and three proximal androgen-response elements (AREs) that physically interact with the DNA-binding domain of AR (Qi, Gao and Wang, 2008). This evidence identifies molecular targeting of the androgen/TGF-β signaling cross-talk as an attractive therapeutic approach in prostate cancer.

The TGF-β family members not only regulate proliferation, growth arrest, and apoptosis of prostate stromal and epithelial cells, but also the formation of osteoblastic metastases. TGF-β is overexpressed in advanced prostate cancer, and exerts diverse functions in stromal cells via both Smad-dependent and Smad-independent signaling pathways (Roberts and Sporn, 1986; Derynck and Zhang, 2003; Zhu and Kyprianou, 2005; Zhu *et al.*, 2006). Cancer cells become refractory to the growth inhibitory activity of TGF-β due to loss or mutation of transmembrane receptors or intracellular TGF-β signaling effectors during tumor initiation (Akhurst and Derynck, 2001). During prostate tumor progression to metastatic disease, TGF-β1 ligand overexpression results in pro-oncogenic rather than growth-suppressive effects. In human prostate cancer cells, interaction of Smad4 (alone or together with Smad3) with the AR in the DNA-binding and ligand-binding domains, may result in the modulation of DHT-induced AR transactivation (Zhu *et al.*, 2008). Interestingly, in the human prostate cancer cell lines PC-3 and LNCaP, addition of Smad3 enhances AR transactivation, while co-transfection of Smad3 and Smad4 actually repress AR transactivation (Kang *et al.*, 2002). A protein–protein interaction between AR and Smad3 has been documented both *in vitro* and *in vivo* via the transcription activation domain of AR and the MH2 of Smad3; AR repression by Smad3 is mediated through the MH2 domain (Hayes *et al.*, 2001). In PC-3 prostate cancer cells, AR expression reduces the TGF-β1/Smad transcriptional activity and the growth effects of TGF-β1, thus preventing TGF-β1 induced apoptosis. Furthermore, TGF-β1 suppresses the E2F transcriptional activity of AR activation, an action associated with a reduced c-Myc expression. An ARE sequence in the TGF-β1 promoter may provide a mechanistic basis for TGF-β1 promoter activity towards DHT in both Huh7 and PC3/AR expressing cells. A direct interaction between AR and TGF-β1 has been causally implicated in other human tumors, including hepatocarcinogenesis (Yoon *et al.*, 2006). Androgens can inhibit TGF-β1-induced transcriptional activity in prostate cancer cells (Chipuk *et al.*, 2002), an interaction that is regulated by AR-associated protein 55 (ARA55/Hic-5) (LIM protein superfamily). Overexpression of ARA55 inhibits TGF-β-mediated up-regulation of Smad transcriptional activity in rat prostate epithelial cells, as well as in human prostate cells, via an interaction between

ARA55 and Smad3 mediated through the MH2 domain of Smad3 and the C terminus of ARA55 (Wang *et al.*, 2005).

Work performed by our group demonstrated the contribution of AR in TGF-β apoptotic signaling in prostate cancer cells. Treatment of the androgen-sensitive TGF-β receptor II overexpressing LNCaP TβRII cells with TGF-β in the presence of androgen led to a significant apoptosis induction via caspase-1 activation and BCL-2 targeting (Bruckheimer and Kyprianou, 2001). Enforced BCL-2 expression significantly inhibits the combined TGF-β and DHT apoptotic effect in prostate cancer cells (Bruckheimer and Kyprianou, 2002). A potential androgenic contribution, with TGF-β enhancement, on EMT, provides an attractive mechanistic possibility in view of the assigned role of EMT during cancer metastasis (Zavadil and Bottinger, 2005), with E-cadherin being considered as an attractive target for such a dynamic duo.

TGF-β causes cell cycle arrest in the G1 phase by up-regulating certain CDK inhibitors such as p15, p21, and p27 (Guo and Kyprianou, 1998; Kyprianou, Bruckheimer and Guo, 2000). During normal development, TGF-β restrains proliferation through induction of cytostatic and apoptotic gene programs; the transcriptional driving element, the TGF-β inducible early-response gene (TIEG1), leads to inhibition of cell proliferation in epithelial cells (Siegel and Massague, 2003). Early *in vivo*-based evidence revealed that exogenous administration of TGF-β into the rat ventral prostate leads to apoptosis of glandular epithelial prostate cells (Martikainen *et al.*, 1991). Proteomic-based studies have identified two proteins, cofilin and prohibitin, as novel mediators of the androgen and TGF-β signaling involved in cell proliferation and cell death functions in prostate cancer cells (Gamble *et al.*, 2004; Zhu *et al.*, 2006; Zhu *et al.*, 2010a). These findings are mechanistically intriguing, as they indicate that changes in the expression and topological distribution of cofilin and prohibitin contribute to the apoptotic and migration regulatory effects of TGF-β1 in prostate cancer cells, providing a molecular basis for its nature as a double-edged sword (Akhurst and Derynck, 2001; Manjeshwar *et al.*, 2003). But Smad7 as an important target for TGF-β-induced apoptosis cannot be ignored. By engaging p53 and p38, Smad7 fulfills the apoptotic function in several tumor cell types, including prostate cancer cells (Edlund *et al.*, 2003; Zhang *et al.*, 2004b; Edlund *et al.*, 2005). Moreover, the TNF receptor associated factor 6 (TRAF6) carboxyl homology domain interacts with TGF-β receptors, and TRAF6 is specifically required for JNK and p38 pathways towards induction of both apoptosis and EMT in response to TGF-β (Zhan *et al.*, 2004; Sorrentino *et al.*, 2008).

In normal prostate epithelial cells and pre-malignant lesions (high-grade intraepithelial neoplasia), TGF-β plays a positive role as a tumor suppressor; thus targeting TGF-β too early in the tumorigenic process might generate the possibility of preventing its tumor-suppressing activity, which would enable rapid tumor growth. The intricate details of the TGF-β transduction network that signal growth-inhibitory versus growth-promoting responses has been extensively investigated (Bachman and Park, 2005). Members of the TGF-β signaling family are emerging as promising predictive biomarkers and molecular targets for the prevention and treatment of metastatic prostate cancer (Jakowlew, 2006; Jones, Pu and Kyprianou, 2008). Indeed, elevated TGF-β levels in a patient's serum provides a significant prognostic value for highly aggressive metastatic disease and poor patient prognosis (Biswas *et al.*, 2006).

In advanced prostate cancer, there is elevated expression of TGF-β with a simultaneous loss of expression of TβRI and TβRII receptors. TGF-β overproduction enhances the metastatic ability of prostate tumor cells and provides a driving mechanism for tumor vascularity (Wikstrom *et al.*, 1998). Both TβRI and TβRII receptors are required to maintain the functional integrity of TGF-β signaling, and loss of either of these receptors results in resistance to TGF-β, permitting the cells to escape the growth constraints imposed by TGF-β. During tumor progression, TGF-β contributes to the metastatic process by promoting ECM degradation and invasion, and it also affects immunity (Akhurst and Dernyck, 2001; Thomas and Massague, 2005). Malignant cells become resistant to the TGF-β growth-inhibitory effects, potentially via mutations in cell cycle regulators such as p15 (Moutspoulos, Wen and Wahl, 2008), or alterations in the activation of Smad-independent pathways, such as PI3K and Ras, with the Smad-dependent signaling having a gradually diminishing role in the TGF-β growth regulatory network (Park *et al.*, 2003; Derynck and Zhang, 2003). Significantly enough, investigation of biomarkers such as TGF-β level, its signaling regulators, effectors, and receptors in prostate cancer progression can provide an effective approach for drug targeting.

TGF-β exerts its influence on the tumor microenvironment through multiple means, such as stroma–epithelial interactions (Beck, Schreiber and Rowley, 2001) and by engaging endothelial cell activation towards increased vascularity (Barrett *et al.*, 2006), thus promoting tumor progression to metastasis. The TGF-β family proteins can induce cells that have acquired cell differentiation characteristics of one particular lineage to switch to differentiation along another lineage though paracrine and autocrine signaling mechanisms (Akhurst and Dernyck, 2001). EMT induction by prostate tumor cells represents another functional contribution by TGF-β to tumor progression and metastasis (Tomita *et al.*, 2000; Teicher, 2001; Tahir *et al.*, 2001; Zhu and Kyprianou, 2010). EMT contributes to the generation of various tissues and organs by allowing the differentiation of highly organized, tightly connected epithelial cells to disorganized, motile mesenchymal cells that phenotypically resemble fibroblasts (Thiery, 2003). Cells undergoing EMT display a loss of cell-to-cell adhesion properties, characterized by decreased E-cadherin, ZO-1, and integrin β1 expression (Zavadil and Bottinger, 2005). Such loss of cell adhesion proteins ultimately allows the cells to be digested and to migrate through the ECM. Induction of an

EMT phenotype in prostate tumor epithelial cells is impaired by blocking the PI3K-AKT pathway, thus implicating non-Smad pathways in the EMT response of prostate cancer cells (Ao *et al.*, 2006). The biological consequences of TGF-β-directed EMT are of major significance in dissecting the mechanisms driving prostate cancer metastasis; TGF-β takes a major role in the regulation of bone metastasis of prostate cancer cells. Another member of the TGF-β superfamily, bone morphogenetic protein 7 (BMP7), which directly interacts with TGF-β, is critical in the regulation of EMT causally linked to bone metastasis of prostate cancer cells (Buijs *et al.*, 2007).

Stromal–epithelial interactions comprise a dynamic tumor microenvironment during prostate cancer progression. Indeed, carcinoma-associated fibroblasts (CAF) derived from prostate tumors can stimulate tumor progression of non-transformed prostate epithelial cells (Hayward *et al.*, 1998). Growth factors expressed by stoma/fibroblasts can exert a paracrine growth influence by binding to receptors on adjacent epithelial cells, or they can exert an autocrine influence by binding to receptors on other stromal cells. Elegant *in vivo* studies (Bhowmick *et al.*, 2004) demonstrated that in a mouse model, inactivation of the TβRII gene in mouse fibroblasts disrupted the epithelial–mesenchymal interaction, resulting in intraepithelial neoplasia in the mouse prostate. Furthermore, hepatocyte growth factor (HGF) signaling was highly activated in the prostate of inactivated TβRII mouse. Thus, epithelial cells are stimulated to release growth factors such as TGF-β that induce stromal cell growth, allowing a robust dynamic of stromal–epithelial interactions during prostate tumor development and progression (Reynolds and Kyprianou, 2006). Increased levels of TGF-β directly correlate with the capacity of CAF to promote the transformation of epithelial cells. Moreover, when human androgen-sensitive prostate cancer cells (LNCaP cells) are co-transplanted into mouse tumors with CAF and ECM components, the resulting tumors exhibit a tenfold higher micro-vessel density than in the absence of CAFs (Ao *et al.*, 2006). The ability of TGF-β to regulate the reactive stroma and promote angiogenesis during progression to metastasis is strongly established (Tuxhorn *et al.*, 2002a).

### 4.4.4 *AR and FGF interactions*

The fibroblast growth factor (FGF) family is a large family of proteins with broad spectrum of functions, including cell migration, differentiation, and angiogenesis (Ornitz and Itoh, 2001). Changes in the expression of FGFs and/or their receptors are involved in prostate tumor progression towards androgen-independent disease (Dow and deVere White, 2000). The estrogen receptor (ER) can regulate the synthesis of FGF-2 and FGF-7 in prostate cells, while stromal ER can mediate the synthesis of stromally derived growth factors, both in coordination with AR activation. AR signaling can directly dictate dramatic changes in the expression pattern of FGFs in both prostate tumor epithelial cells and stromal cells, primarily via changes in FGF-1, FGF-2, FGF-8, and FGF-10 (Nakano *et al.*, 1999; Saric and Shain, 1998; Rosini *et al.*, 2002). AR is upregulated by paracrine FGF-10 via a positive feedback, and synergizes with cell-autonomous activated AKT in prostate cancer cells (Memarzadeh *et al.*, 2007). Moreover, in response to FGFs, AR facilitates FGF-induced survival of prostate cancer cells, possibly through BCL-2 induction and down-regulation of AR, allowing the escape of selected clones from androgenic control (Rosini *et al.*, 2002; Gonzalez-Herrera *et al.*, 2006).

### 4.4.5 *AR and VEGF: Vascular exchanges for the "road"*

Vascular endothelial growth factor (VEGF), originally known as vascular permeability factor, is a well-characterized angiogenic cytokine, responsible for endothelial cell proliferation, migration, and vessel assembly (Millauer *et al.*, 1993; Fong *et al.*, 1995). Its value as a diagnostic tool as well as a therapeutic target for advanced metastatic prostate cancer has been examined at the molecular and translational level (Borgstrom *et al.*, 1998). The "hypoxia-response" signaling system upregulates the expression of a network of effectors that increases the propensity of tumor cells for survival, even in this adverse environment (Anastasiadis *et al.*, 2003; Ghafar *et al.*, 2003). Expression of VEGF is transcriptionally induced by hypoxia-inducible factor (HIF-1$\alpha$) in response to oxygen changes in the microenvironment (Delongchamps, Peyromaure and Dinh-Xuan, 2006). Androgen-stimulated growth of the glandular ventral prostate is preceded by increased VEGF synthesis, endothelial cell proliferation, vascular growth, and increased blood flow (Joseph *et al.*, 1997; Franck-Lissbrant *et al.*, 1998). The role of VEGF in androgen-mediated prostate vascularity was further supported by additional studies (Lissbrant *et al.*, 2004). In prostate cancer, the effect of androgens on angiogenesis is mediated via their ability to regulate VEGF through activation of HIF-1$\alpha$ in androgen-sensitive tumors (Boddy *et al.*, 2005). The significant correlation between HIF-1$\alpha$ and HIF-2$\alpha$ expression and with AR and VEGF expression (Boddy *et al.*, 2005; Banham *et al.*, 2007) provides firm support for such a control system. The driving mechanism involves the direct up-regulation of VEGF-C in response to androgen depletion in prostate cancer cells, via activation of the small GTPase, RalA; VEGF-C can increase the AR co-activator BAG-1L expression that facilitates AR transactivation. Under conditions of low androgen levels, the intracellular reactive oxygen species induce RalA activation and VEGF-C synthesis (Rinaldo *et al.*, 2007).

### 4.4.6 *AR and growth factor interplay in the stroma*

The stroma is a lead component of the prostate microenvironment contributing to tumor heterogeneity and growth dynamics. Stroma-derived fibroblasts play an active role in carcinogenesis in addition to structurally supporting the epithelial cell growth (Chung *et al.*, 1989; Camps *et al.*, 1990; Chung *et al.*, 1991; Cunha *et al.*, 1996). Studies in the early 1990s, pioneered by Leland Chung's group, established that human prostate-derived stromal cells stimulate growth of prostate cancer epithelial cells *in vitro* and *in vivo*, and such reciprocal interactions can dictate hormonal responsiveness (Gleave *et al.*, 1991; Chung *et al.*, 1991). This evidence widely popularized the belief that disturbance in the epithelial–stromal interactions is most critical in the pathogenesis of prostate cancer and progression to an androgen-independent (castration-resistant state) (Hayward *et al.*, 1998). Androgenic control during normal growth and differentiation of the prostate gland is regulated via nuclear AR in both stromal and epithelial cells (Sar *et al.*, 1990). The close association between low AR levels in the stroma, adjacent to malignant epithelium with a poor clinical outcome in prostate cancer patients, is of high translational value (Henshall *et al.*, 2001). Androgens increase VEGF transcription and active VEGF secretion from prostatic stroma, thus indirectly enhancing prostate cancer growth and angiogenesis (Levine *et al.*, 1998). DHT and bFGF can synergistically stimulate prostate stromal cell proliferation (Niu *et al.*, 2001), while androgen depletion rapidly reduces stroma IGF-1 synthesis and its action in the prostate epithelium. Close rules of compartmentalization become "loose" here: although IGF-1 is principally produced in the stroma and IGF-R1 in the epithelium, both are under androgenic regulation as stroma IGF-1 mRNA is significantly decreased after castration, correlating with epithelial cell apoptosis (Ohlson, *et al.*, 2007).

TGF-β1 is also regulator of stromal cell proliferation and differentiation, depending on the specific stromal cell type, microenvironment, and contributing activities of other growth factors (Sporn and Roberts, 1992). A distinction in the complex cross-talk between androgens and TGF-β1 signaling in prostate stromal cells, affects AR localization, cell proliferation, and myodifferentiation, thus defining its mechanistic contribution to the reactive stroma. AR and TGF-β1 levels significantly correlate in the stromal component of prostatic intraepithelial neoplasia (PIN) (Cardillo *et al.*, 2000). Induction of rat PS-1 prostate stromal cell proliferation by androgens can be antagonized by TGF-β1. Furthermore, TGF-β1 triggers a cytoplasmic translocation of nuclear AR during myodifferentiation in the prostate stroma (Gerdes *et al.*, 1998; Gerdes *et al.*, 2004), while androgens enhance TGFβ1-mediated proliferation of prostatic smooth muscle cells PSMC1 (Salm *et al.*, 2000).

During prostate cancer progression the androgen axis engages the growth factor network to an active cross-talk towards conferring a survival and invasion advantage of prostate cancer cells. Androgens can modify prostate cancer cell response to growth factor signals, from growth-inhibitory, to tumor-promoting, during the metastatic process. A better understanding of such cross-talk between the AR axis and critical growth factor signaling in the context of the tumor microenvironment may identify a mechanism underlying the emergence of androgen-independent prostate cancer, and provide new opportunities for therapeutic targeting of aggressive prostate tumors.

# 5

# Anoikis in Prostate Cancer
# Metastasis

## 5.1 Anoikis Interrupted: Survival of the Homeless (Cells)

The emergence of disseminated metastasis remains the primary cause of mortality in cancer patients. The ability of cancer cells to migrate is universal to all human malignancies, an aberrant behavior manifested as invasion (the ability of tumor cells to become dislodged and travel within the tissue of origin) and metastasis (involving travel beyond the primary tumor origin and establishing a niche at a distant site). While sharing some molecular pathways with the other characteristic feature of tumor cells — increased proliferation and reduced apoptosis — inappropriate cell motility is a distinct process that can make malignant epithelial cells a "moving target". The engagement of a common set of factors in distinct steps of metastasis qualifies them as metastasis progression genes, functionally involved in metastatic colonization (Brabletz *et al.*, 2001; Yoshida *et al.*, 2001) The attachment of epithelial and endothelial cells to the ECM is essential for the tight regulation of cell proliferation, cell survival, and migration. The functional contribution of integrins is essential for such processes, as integrins recognize the major adhesive ECM components, fibronectin and laminin (Giancotti and Ruoslahti, 1999; Goel and Languino, 2004). Integrins are transmembrane proteins that serve a role as primary mediators of cell-ECM interactions that are functionally involved in determining tumor angiogenic response during cancer progression to metastatic disease (Fornaro, Manes and Languino, 2001; Goel and Languino, 2004; Goel *et al.*, 2008). Efforts to identify genetic determinants of metastasis led to gene "signatures" for which the expression in primary tumors is associated with high risk of metastasis and poor survival (Fig. 3.3). Anoikis, a Greek word meaning "loss of home" or "homelessness" was originally defined by Frisch a decade ago as a unique phenomenon reflecting a specific mode of apoptosis induced by inadequate or inappropriate cell-matrix interactions (Frisch and Ruoslahti, 1997; Frisch and Screaton, 2001) and is a potentially significant player in angiogenesis and metastasis during tumorigenesis (Rennebeck, Martelli and Kyprianou, 2005). As a functional phenomenon, anoikis suppresses expansion of oncogenically transformed cells by preventing proliferation at distant locations, while migrating tumor cells that are resistant to anoikis induction can grow at inappropriate locations (Frisch and Francis, 1994; Debnath *et al.*, 2002). Once within the circulation, cells with high metastatic potential can evade immune surveillance, invade distal organs, and initiate *de novo* tumor growth. Thus resistance to anoikis enables malignant cells to survive in an anchorage-independent manner which leads to the formation of distant metastasis. The central problem of the anoikis killing versus survival effect is to understand how integrin-mediated cell adhesion signals control the apoptotic machinery (Fukuda *et al.*, 2003). Tumor cells can escape from detachment-induced apoptosis by controlling anoikis pathways; molecules such as galectin-3 are recruited to dictate the outcomes of the ECM integrin cell survival pathway, as well as the extrinsic death receptor pathway (Oka, Takenaka and Raz, 2004; Nakahara, Oka and Raz, 2005), once the anoikis command is received (Rabinovitz, Gipson and Mercurio, 2001). The very ability of the cell to survive in the absence of adhesion to the ECM enables tumor cells to disseminate from the primary tumor site, invade a distant site, and establish a metastatic lesion. It is not surprising therefore that resistance to anoikis is quickly emerging as a hallmark feature of metastatic cancer cells, including prostate cancer cells (Duxbury *et al.*, 2004; Giannoni *et al.*, 2008; Sakamoto and Kyprianou, 2010). Consequently, identification of pathways that regulate anchorage-independent survival is fundamentally important in dissecting the process of metastasis, and may provide novel leads for therapeutic strategies specifically targeting metastatic disease.

Androgen-independent prostate cancer cells become resistant to apoptosis-activating agents due to roadblocks in apoptosis, and, depending on the interactions with the tumor microenvironment, acquire invasive and metastatic properties towards the castration-resistant androgen-independent state (Nasu *et al.*, 1998). Understanding the molecular mechanisms rendering prostate tumor cells resistant to anoikis (activation of survival signals that are not ECM contact-dependent and/or inhibition of apoptotic pathways) will enable the targeting of new routes of prostate cancer metastasis. It is an exploitation that holds considerable therapeutic promise, as it may effectively lead to the identification of events that can be manipulated selectively for the metastatic cancer cells:

1. reversing the ability of tumor cells to become resistant to anoikis, therefore making them more susceptible to anoikis-inducing agents;

2. interfering with the seeding process of tumor cells into secondary places, by making tumor cells non-sensitive to the chemotactic and environmental cues of the new target organ; and
3. making these secondary targets less "appealing" to the cancer cells by blocking key molecules promoting cancer cell seeding and survival.

The evidence so far implies that circumvention of anoikis may be involved on a more global scale in human cancer. There are two primary mechanistic platforms on which anoikis resistance can make its functional contribution to the process of prostate cancer progression to metastasis:

- During reactive tumor stroma formation.
- In epithelial–mesenchymal transition.

The presence of circulating cancer cells after adjuvant chemotherapy is prognostic for an increased risk of relapse and the development of prostate cancer metastatic disease (Fleisher *et al.*, 2009). True to the notion "live free or die", in the context of the circulating "homeless" cells, the anoikis phenomenon can effectively block metastasis by preventing normally adherent cells from surviving after detachment from the primary site (Berezovskaya *et al.*, 2005). A novel high throughput screen reported by Mawji and colleagues identified anisomycin as a potent anoikis sensitizer, triggering anoikis by decreasing FLIP protein synthesis and activating the death receptor pathway, totally independent of its ability to activate c-Jun-MH2-kinase (JNK) and p38 (Mawji *et al.*, 2007). Significantly enough, consistent with its functional promotion of anoikis, anisomycin also reduced the metastatic potential of human prostate cancer cells (by specifically targeting their survival) in a mouse model of prostate cancer metastasis.

## 5.2 The Integrin Connection

Cell adhesion to the ECM operates during development to ensure proper tissue organization and cell migration. Integrins are transmembrane proteins representing the main class of receptors for ECM proteins, such as collagen and laminin found in basement membranes (Goel and Languino, 2004). These cell-surface proteins serve a role as primary mediators of cell-ECM interactions that are functionally involved in determining tumor angiogenic response and metastatic progression. Cell adhesion to the ECM causes reorganization of the actin cytoskeleton and disruption of actin polymerization impairs cell adhesion (Giancotti and Ruoslahti, 1999). Biochemical evidence indicates that mechanotransduction through integrins involves physical linkage of integrins and cytoskeletal proteins (Fornaro, Manes and Languino, 2001; Goel *et al.*, 2008). Moreover, growth factor signaling events and cell adhesion mechanisms operate in a cooperative fashion to promote mitogenesis in adherent cells, and implicate integrin-mediated cell adhesion in generating discrete growth-regulatory cues (Kornberg and Liberti, 1992). Angiogenic growth factors such as VEGF and pFGF can exert a profound positive effect on the activity and expression of several integrins, such as $\alpha_v\beta_3$, $\alpha_v\beta_5$, $\alpha_v\beta_1$, $\alpha_3\beta_1$, $\alpha_6\beta_1$, and $\alpha_6\beta_4$ (Klein *et al.*, 1993; Holash *et al.*, 2002). Additional angiogenic factors, such as class 3 semaphorins (SEMA3 proteins), control vascular remodeling via targeting integrin (Serini *et al.*, 2003). Increased expression of integrin $\alpha_v\beta_3$ is detected in growth factor-activated endothelial cells in tumor blood vessels and granulation tissue (Enenstein, Waleh, and Kramer, 1992; Brooks *et al.*, 1994), while very low expression was detected in resting blood vessels (Arap *et al.*, 2002). Expression of the $\alpha_v\beta_3$ in tumor endothelium correlated with the aggressive phenotype in neuroblastoma (Erdreich-Epstein *et al.*, 2000). Moreover, blocking integrin $\alpha_v\beta_3$ with monoclonal antibody mediated endothelial cell apoptosis, and inhibited blood vessel formation (Enenstein, Waleh, and Kramer, 1992), implicating its functional significance in angiogenesis. Consequently, $\alpha_v\beta_3$ integrin represents an attractive tumor vasculature target (Ruoslahti, 2002; Hood and Cheresh, 2002). Nanoparticles containing agents linked with an anti-$\alpha_v\beta_3$ antibody (Sipkins *et al.*, 1998) or bacteriophage with $\alpha_v\beta_3$ binding RGD peptide (Arap *et al.*, 2002) effectively targeted tumor vasculature. Integrins also contribute to signal transduction from the extracellular environment to the intracellular network, mediated by integrin-activated signaling molecules, such as focal adhesion kinase (FAK) and phosphatidylinositol 3-kinase (PI 3-kinase) (Frisch *et al.*, 1996), members of the extracellular signal-regulated kinase 1 and 2/mitogen activated protein (ERK1 and 2/MAP) kinase family, as well as the guardian of the genome, p53 (Zhang *et al.*, 2004b). The non-receptor tyrosine kinase FAK is a makor mediator of integrin signaling, since engagement of beta1 and av (but not b4) integrins induces FAK localization to matrix adhesions and activation. Upon autophosphorylation Tyr 397, FAK combines with Src or SFK to phosphorylate FAK-associated focal adhesion proteins, such as paxillin or tyrosine residues. In addition FAK can combine with PI3K to activate the tyrosine kinase ERK (Fig. 3.3). Through this plethora of interactions with downstream effectors, FAK regulates focal adhesion dynamics and activates proliferation, migration, survival, and anoikis signaling pathways in tumor epithelial and endothelial cells (Fornaro, Manes and Languino, 2001; Nikolopoulos *et al.*, 2004).

Clustering of β1-integrin in the absence of growth factors stimulates rapid and transient tyrosine phosphorylation of intracellular proteins, leading to activation of kinase signaling involved in cancer development (Goel and Languino, 2004). The dynamic of the ECM–cell interaction provides cells the information regarding its context within a tissue that can control cell proliferation, migration, differentiation, and survival. Fibroblasts, endothelial, and epithelial cells are dependent upon the ECM for survival and proliferation. Disruption of the cell-matrix contacts can trigger anoikis in all three cell types (epithelial, endothelial, and fibroblast) and ultimately leading to caspase-dependent apoptosis within minutes, upon detachment (Frisch and Ruoslahti, 1997; Aoudjit and Vuori, 2001; Loza-Coll *et al.*, 2005).

Development of prostate cancer bone metastasis is a multistep process that initially involves loss of cell–cell contact and intravasation through the basement membrane. The tumor cells then circulate through the blood or lymphatic system, while evading the immune system, eventually being carried to the bone marrow. Once the tumor epithelial cells invade through the endothelial lining of blood vessels they interact with cells in the bone microenvironment to become established metastatic lesions (Rennebeck, Martelli and Kyprianou, 2005; Sakamoto *et al.*, 2010).

## 5.3 Impairing the Route to Angiogenesis

Neovascularization of tumors is a critical process tightly linked to reactive stroma and cancer progression. Historically the first recognition of the significance of vascularization in cancer was reported 65 years ago by Glenn Algire, who demonstrated that the rapid growth of tumor explants was dependent on the development of a rich vascular supply (Algire and Chalkley, 1945). Fast-forward 25 years to the pioneering work of Judah Folkman that defined tumor angiogenesis as an integral part of tumorigenesis, and the therapeutic potential of targeting tumor angiogenesis for cancer treatment (Folkman, 1971). Since then anti-angiogenic therapies are being developed at phenomenal pace and translated to the clinic with great promise (Folkman, 1995b; Albig and Schiemann, 2004); such clinical efforts, however, have met with questionable success possibly due to mechanisms of resistance to anti-angiogenic drugs (Bergers and Hanahan, 2008). But is angiogenesis linked to the apoptosis routing? In normal tissue homeostasis a fine balance between angiogenesis promoters and angiogenic blockers controls angiogenesis and tissue vascularity (Dhanabal *et al.*, 1999; Bergers and Benjamin, 2003). Tumors, like normal tissues, have physiological constraints on growth, such as access to oxygen and nutrients for metabolism; thus tissue growth is restricted to a few cubic millimeters if no new vasculature is formed, and tumors remain in a dormant state unless they are able to recruit their own vasculature (Folkman, 1993). This size restriction on non-vascularized tumors is due to the high rate of apoptosis resulting from the inadequate supply of oxygen. Advanced tumor growth is critically dependent upon excessive angiogenesis towards development of a new blood supply (Greenblatt and Shubick, 1968; Carmeliet, 2000). This is accomplished by the massive production of angiogenic factors, termed the angiogenic switch, secreted into the local tissues and stroma (Hanahan and Folkman,

1996). This new vascularization not only enhances tumor growth, but also provides a route for metastatic spread, which clinically is the most lethal aspect of cancer and the leading cause of cancer-related deaths. The ECM remodeling provides a spatial stage for the release of growth factors that promote angiogenesis, while nurturing their interaction with other signaling events depending on the cell context (Liotta, Steeg and Stettler-Stevenson, 1991; Gullberg *et al.*, 1992; Fong *et al.*, 1999). Thus FGF2 stimulates fibroblast proliferation, ECM production, protease secretion, and also functions as an angiogenic factor by inducing migration of endothelial cells and supporting their differentiation into new blood vessels (Lyden *al.*, 2001). Vascular endothelial growth factor (VEGF), perhaps the most potent angiogenesis promoting factor, is fundamentally critical in embryonic development (Ferrara *et al.*, 1996) and is upregulated in numerous human tumors (Ferrara *et al.*, 2003). Blocking the activity of VEGF through neutralizing antibodies abrogates angiogenesis and prevents the metastatic spread of several human malignancies (Boehm *et al.*, 1997; Presta *et al.*, 1997; Mordenti *et al.*, 1999; Holash *et al.*, 2002; Hu *et al.*, 2002).

The emerging therapeutic value in prostate cancer of matrix-derived angiogenic inhibitors such as arresten, canstatin, endorepellin, endostatin, anastellin, fibullins, and thrombospondin commands attention (Dhanabal *et al.*, 1999; Marneros and Olsen, 2001; Nyberg, Xie and Kalluri, 2005; Horsman and Siemann, 2006). Endostatin is a 20kDa fragment from the COOH-terminal portion of the NC-1 domain of type XVIII collagen, which blocks angiogenesis without any associated toxicity or resultant drug resistance (O'Reilly *et al.*, 1997). Endostatin binds directly to the VEGF receptor (Abdollahi *et al.*, 2004), thus preventing endothelial cell migration and inducing apoptosis (Klement *et al.*, 2000; Huss, Barrios and Greenberg, 2003). Endostatin can also regulate endothelial cell adhesion and cytoskeletal organization (Kim *et al.*, 2000; Dixelius *et al.*, 2002), possibly via its association with the integrin $\alpha5\beta1$ towards Src activation by recruiting a tyrosyl phosphatase-dependent pathway (Wickstrom, Alitalo and Keski-Oja, 2002). In addition endostatin can interfere with the FGF signaling towards abrogation of MMP activation, via direct binding to MMP2 and MMP9 (Klein *et al.*, 1993; Kleiner and Stetler-Stevenson, 1999; Kim *et al.*, 2000; Lee *et al.*, 2002b; Hiratsuka *et al.*, 2002; Nyberg *et al.*, 2003).

It is well recognized that in the prostate gland, cells residing towards the lumens in multilayered cribriform and/or solid intra-glandular neoplastic lesions have decreased access to blood vessels and, therefore, decreased access to both nutrients and oxygen. Androgens normally stimulate the production of vascular regulatory factors by stromal cells, including VEGF and angiopoietins (Jackson, Bentel and Tilley, 1997), the absence of which post castration leads initially to apoptotic loss of endothelial cells and vasoconstriction of the remaining blood vessels, followed temporally by apoptosis of prostate glandular epithelial cells and tissue regression (Joseph *et al.*, 1997; Hayek *et al.*, 1999; Buttyan, Ghafar and Shabsigh, 2000). In the experimental model of prostate adenocarcinoma Dunning R3327, castration-induced androgen deprivation leads to a dramatic loss of both VEGF and its receptors (Haggstorm *et al.*, 1998).

Thrombospondin-1 (TSP-1) is a homotrimeric multidomain glycoprotein that has been found to correlate with non-metastatic tumor phenotypes, though it may also enhance angiogenesis by modulating cell adhesion and survival in certain cellular settings (Lawler *et al.*, 2000; Jin *et al.*, 2000; Doll *et al.*, 2001; Urquidi *et al.*, 2002). This may be a consequence of thrombospondin's ability to activate TGF-$\beta$, a key multifunctional growth factor that contributes to tumorigenesis by regulating apoptosis, proliferation, and ECM degradation (Roberts, 1996). Studies on tumor transplantation into TSP-1 null mice demonstrated that tumor cell proliferation is increased and tumor apoptosis is decreased (Holmgren, O'Reilly and Folkman, 1995). An intriguing feature of thrombospondin is its ability to distinguish between pathological neovascularization and normal pre-existing vascularization (Noh *et al.*, 2003) by engaging the FAS/FAS-L-mediated apoptotic signaling, since FAS-L receptor expression is relatively low in quiescent endothelial cells but rather high in proliferating endothelial cells (Rodriguez-Manzaneque *et al.*, 2001; Volpert *et al.*, 2002; Veitonmaki *et al.*, 2004).

Angiostatin is another potent endogenous inhibitor of angiogenesis, often detected in tumor tissues in association with tumor growth (Carmeliet, 2000). Angiostatin significantly reduces tumor volume, mass, and microvessel density in the human hepatocellular carcinoma (Tao, Dou and Wu, 2004). Production of angiostatin in athymic nude mice bearing MDA-MB-435 xenografts results in massive endothelial cell apoptosis and vascular collapse within the tumor core, leading to significant tumor cell apoptosis/necrosis (Agarwal *et al.*, 2004). The *in vivo* data from human prostate cancer xenografts in athymic mice demonstrated that a combined treatment of angiostatin and docetaxel (a taxane) led to significant tumor regression due to a marked reduction in intratumoral vascularization (Galaup *et al.*, 2003).

Tumor cell repopulation after certain chemotherapeutic approaches, using vascular disrupting agents, may be assisted by a reactive host response involving the mobilization of bone marrow-derived circulating cells, which subsequently home to the vasculature of treated tumors, thus promoting further neovascularization. Such vasculogenic "rebounds" may be effectively blocked at their initiation by exposure of tumors to anti-angiogenic drugs such as quinazoline-based doxazosin. Maspin (mammary serine protease inhibitor), a member of the serpin family, is a tumor-suppressor protein that regulates cell migration, invasion, and adhesion (Sager *et al.*, 1994). Maspin is down-regulated in epithelial cells from invasive prostate tumors serving as an angiogenesis inhibitor (Zhang *et al.*, 2000). Maspin expression in prostate cells is under hormonal regulation mediated by the AR (Zhang *et al.*, 1997). An interaction between cell surface maspin and key components of the ECM (such as collagen), potentially enhanced under hypoxic conditions, may drive its ability to enhance cell adhesion and inhibit cellular migration of prostate tumor epithelial and endothelial cells *in vitro* and *in vivo* (McKenzie, Sakamoto and Kyprianou, 2008). The "appeal" of the therapeutic value of maspin stems from two lines of pharmacologic evidence:

- Targeted expression of maspin in tumor vasculature induces massive epithelial cell apoptosis (Li, Shi and Zhang, 2005), possibly via the mitochondrial death pathway (Liu *et al.*, 2004)
- Maspin sensitizes human prostate tumor cells to doxazosin-mediated anoikis (a quinazoline-based $\alpha_1$-adrenoceptor antagonist) (Tahmatzopoulos, Sheng and Kyprianou, 2005), while its systemic administration has a very low toxicity profile.

Angiogenesis-targeted therapies offer an alternative management perspective in prostate cancer, attractive because of its indolent clinical course and low proliferative capacity (Lara, Przemyslaw and Quinn, 2004; Sakamoto, Ryan and Kyprianou, 2008). Several clinical trials using potent anti-angiogenic drugs able to trigger apoptosis selectively among the tumor-associated endothelial cells have been undertaken in prostate cancer patients, with great expectations. These expectations, however, have not been completely realized. The anti-angiogenic agents including TNP-470 (Logothetis *et al.*, 2001), 2-methoxyestradiol (Sweeney *et al.*, 2005), SU11248 (Mende *et al.*, 2003) and SU5416 (Stadler *et al.*, 2004), failed to deliver any positive therapeutic efficacy in prostate cancer patients with CRPC. Despite the disappointing clinical outcomes, a series of new angiogenesis-targeted approaches have emerged from the laboratory to the clinic, promising considerable clinical benefit.

Selective vasculature targeting has been exploited with great interest and anticipation for prostate cancer therapy (Kopetz *et al.*, 2002). Striking work by Dr Ruoslahti and his colleagues identified the peptides that specifically recognize the vasculature in the prostate through screening phage-displayed peptide libraries. Intravenous injection of chimeric peptide consisting of the SMSIARL homing peptide, linked to a proapoptotic peptide, induced prostate tumor destruction and delayed cancer development in TRAMP mice (Arap *et al.*, 2002).

### 5.3.1 *Doxazosin*

Experimental evidence indicates that the quinazoline-based compounds, such as doxazosin and terazosin (clinically used for treatment of hypertension and BPH), may also function in inducing prostate smooth muscle cell death via apoptotic pathways involving activation of TGF-β1-mediated apoptotic signaling and inhibition of AKT (Kyprianou, 2003; Partin, Anglin and Kyprianou, 2003; Garrison and Kyprianou, 2006). Growing evidence also implicates anoikis (ECM detachment-induced apoptosis) in prostate tumor epithelial and endothelial cells in response to quinazolines (Rennebeck, Martelli and Kyprianou, 2005), potentially via targeting VEGF-mediated angiogenic response (Garrison and Kyprianou, 2006; Tahmatzopoulos and Kyprianou, 2004).

### 5.3.2 Suramin

The therapeutic efficacy of suramin was analyzed in a large population of patients (390) with advanced prostate cancer with a randomized dosing. The objective response rate was 9–15%, while the PSA response rates were 24–34% (Small *et al.*, 2002). In a randomized phase III trial of suramin in combination with hydrocortisone conducted in 460 hormone-refractory prostate cancer patients, patients receiving suramin had a higher PSA response rate, compared to the hydrocortisone arm (Small *et al.*, 2000). There is also experimental evidence to suggest a potential enhancing effect of radiation-induced apoptosis in prostate tumors by suramin (Sklar *et al.*, 1993). Despite the initial excitement about suramin's anti-tumor efficacy in prostate cancer patients there are still serious limitations associated with the clinical efficacy and systemic toxicity of this agent, alone or in combination with other modalities.

### 5.3.3 *Thalidomide*

Pre-clinical experiments revealed that treatment with N-substituted thalidomide — a potent angiogenesis inhibitor and analog CPS11 (targeting platelet-derived growth factor (PDGF) alpha) — led to 90% inhibition of tumor growth and 64% reduction in tumor vascularity of PC-3 human prostate xenografts (Ng *et al.*, 2004). An open trial of the efficacy and safety of thalidomide in 20 patients with androgen-independent prostate cancer resulted in significantly reduced PSA levels in 35% of patients (Drake, Robson, Mehta *et al.*, 2003). Prostate cancer patients with hormone-refractory disease exhibited a relatively good response to a combination treatment of docetaxel and thalidomide, by achieving a significantly improved median overall survival (28.9 months) compared to docetaxel-alone (14.7 months) (Leonard *et al.*, 2003). In a triple-combination strategy of paclitaxel and doxorubicin with thalidomide, 9 out of 12 patients showed a 50% decrease in PSA (Amato and Sarao, 2006). Combinational approaches using granular-macrophage colony-stimulating factor (GM-CSF) have been reported with some therapeutic promise. GM-CSF regulates the dendritic cell and tumor-specific cytokine T-cell mediated response (Small *et al.*, 1999), and clinically, GM-CSF in combination with thalidomide results in a significant PSA response (23%) (Dreicer *et al.*, 2005). Moreover, combination treatment strategies of thalidomide (anti-angiogenesis) and Docetaxel (microtubule-targeting agent), emerge as attractive therapeutic options for castration-resistant non-metastatic prostate cancer (Leonard *et al.*, 2003).

### 5.3.4 *Bevacizumab*

A phase II Cancer and Leukemia Group B (CALGB) 90006 trial of bevacizumab in combination with docetaxel, and estramustine with a pre-medication with decadron in chemotherapy-naïve HRPC delivered promise (with the majority of patients achieving >50% PSA reduction) (Picus *et al.*, 2003) and led to two clinical trials. The ongoing clinical trials are an NCI-phase II study of a four-drug combination strategy of docetaxel, prednisone, thalidomide, and bevacizumab, in men with chemotherapy-naïve progressive hormone-refractory prostate cancer; and a CALGB phase III, double-blind, placebo-controlled trial of docetaxel plus prednisone with or without bevacizumab (ClinicalTrials.gov, 2008).

### 5.3.5 *SU5416*

The effect of a selective small molecule inhibitor of VEGF receptor, vascular endothelial growth factor receptor 2 (VEGFR2 KDR), SU5416, is profound against multiple tumor types, primarily exerted by blocking tyrosine kinase catalysis and impairing tumor vascularization (Fong *et al.*, 1995; Fong *et al.*, 1999). Targeting VEGFR1 in prostate cancer using the transgenic adenocarcinoma of the TRAMP mouse model of prostate cancer progression (Gingrich *et al.*, 1997), and via administration of SU5416 (mice 16–22 weeks of age; highly expressing VEGFR-2), resulted in a dramatic decrease in tumor-associated vascularity and an increase in prostate cancer-specific apoptosis (Huss, Barrios and Greenberg, 2003). Further pre-clinical studies, using PC-3 human prostate cancer xenografts, demonstrated that a combination regime of the VEGFR2 inhibitor, SU5416, with another potent anti-angiogenesis agent, endostatin, led to a significant delay in the onset of tumor progression (Abdollahi *et al.*, 2004).

### 5.4  Anoikis and the Tumor Microenvironment: No "Resting" in the Stroma

The contribution of the stromal layer to the development of a wide variety of tumors has been supported by extensive clinical evidence and by the use of experimental mouse models of cancer pathogenesis (Rowley, 1998; Allinen *et al.*, 2004). Structural differences have been shown between normal and tumor stroma: tumor fibroblasts show enhanced proliferation and migratory activities (Camps *et al.*, 1990; Gleave *et al.*, 1991; Byrne, Leung and Neal, 1996; Tuxhorn,

Ayala, Smith *et al.*, 2002) and there are differences in the constituents of the ECM in tumor versus normal stroma (Chung *et al.*, 1991; Rowley, 1998). Thus the presence of a reactive stroma and EMT are critical steps in the progression of prostate cancer to bone metastasis. Anoikis is structurally dependent on the cellular interactions with the ECM, and with the cytoskeleton rearrangement and organization, and can be regulated by the reactive oxygen species in the microenvironment (Giannoni *et al.*, 2008); one may assume that development of anoikis resistance characteristic in metastatic disease is intimately connected with the above processes and their deregulation. Tumor cells express an array of critical survival factors that functionally contribute towards the metastatic journey to a distant target, in which an appropriately vascularized microenvironment would support their growth and survival. The close functional links between prostate tumorigenesis and loss of apoptosis provides an ever-growing number of molecular insights into the development and clinical formulation of novel apoptosis-targeted cancer therapeutic approaches (Tang and Porter, 1997), as reinstating apoptosis represents a critical defense strategy against chemoresistance of cancer cells (Johnstone, Ruefli and Lowe, 2002; Garrison and Kyprianou, 2004).

The ability of quinazoline-based $\alpha_1$-adrenoceptor antagonists to interfere with anoikis of prostate tumor epithelial cells as well as endothelial cells, via a mechanism targeting VEGF and independent of the $\alpha_1$-adrenoceptor blockade, has been established (Benning and Kyprianou, 2002; Keledjian, Garrison and Kyprianou, 2005). Doxazosin induces morphological changes consistent with anoikis in both benign and malignant prostate cells, by targeting FAK via a BCL-2 independent action, and by FAK levels in PC-3 cells (Keledjian and Kyprianou, 2003). One has to also consider the evidence suggesting that Smad4-induced anoikis may also be dependent upon increased FAK2 protein, a cousin of FAK, and a well-known suppressor of anoikis (Aoudjit and Vuori, 2001). In addition, strong evidence on the therapeutic benefits of induced anoikis stems from experimental studies using human vascular endothelial cells (HUVECs). HUVECs can undergo anoikis when denied matrix attachment and upon activation of the FAS-FASL death receptor-mediated apoptotic pathway (Brooks *et al.*, 1994; Frisch *et al.*, 1996). Rapidly expanding evidence clearly documents that caspase-8 activation and DISC formation are key steps involved in the anoikis-triggering step in human endothelial cells (Keledjian, Garrison and Kyprianou, 2005; Garrison *et al.*, 2007). Cell death in endothelial cells can be blocked upon activation of PI 3'-kinase, the ERK pathway, AKT, or by the broad spectrum caspase inhibitor z-VAD, which may act to prevent the mitochondrial death pathway in HUVECs. Anoikis and caspase-8 activation are inhibited by enhanced expression of c-Flip in detached cells and also by dnFADD and specific inhibitors of caspase-8 (Kurenova *et al.*, 2004; Senft, Helfer and Frisch, 2007). The functional link connecting FAK to apoptosis has not always been linearly defined. While several observations pointed out that FAK is not required for the initial malignant transformation, but rather to promote cancer and endothelial cell invasion, recent evidence suggests that FAK is recruited by Ras signaling to assist in the PI3K-dependent transformation in breast tumorigenesis (Pylayeva *et al.*, 2009). There is also intriguing evidence to implicate FAK overexpression in human tumors in providing a survival signal function by binding to RIP and inhibiting the interaction with the death receptor complex, thus suppressing anoikis (Kurenova *et al.*, 2004).

During androgen ablation, RalA regulates VEGF synthesis in prostate cancer cells, possibly promoting the angiogenic response of CRPC tumors (Rinaldo *et al.*, 2007). Rac1 is a small Rho family GTPase that may prove to have some interesting therapeutic significance in prostate cancer as well. Expression of a constitutively active mutant of Rac1, Rac1-V12, protects Madin-Darby canine kidney (MDCK) cells against anoikis, by securing its adhesion to the integrin beta tail of the ECM (Berrier *et al.*, 2002). Rac1-V12 has also been shown to inhibit anoikis in fibroblasts as well as in a number of untransformed cell lines, implicating a more universal role in the suppression of anoikis.

## 5.5 Signaling the "Homeless" State: Intracellular Anoikis Effectors

Multiple factors with frequently overlapping functions have been reported to confer anoikis suppression to tumor cells, by stimulating survival pathways (Sakamoto and Kyprianou, 2010). One could, however, propose that anoikis resistance not only allows the cell to survive without attachment to the ECM, but it may also provide a molecular signature that secures its successful migration, invasion, and metastasis to a distant site (Frisch *et al.*, 1996; Zhu *et al.*, 2001). Thus key players emerging as novel therapeutic targets might not only participate in the development of anoikis resistance, but might also serve as cellular "ambassadors" of cell migration, seeding, and proliferation of prostate tumor cells to the secondary targets.

Three primary integrin-signaling molecules have been linked to cell survival: focal adhesion kinase (FAK), Src, and integrin-linked kinase (ILK) (Rennebeck, Martelli and Kyprianou, 2005; Diaz-Montero, Wygant and McIntyre, 2006). As illustrated in Fig. 3.1, each of these proteins may impinge upon the PI3K/AKT pathway, but there is also evidence to support signaling via distinct mechanisms. Integrin-mediated cell adhesion stimulates the activity of phosphoinositide-3 kinase (PI3K) and constitutively activated PI3K stimulated ILK, indicating downstream regulation of PI3K activity. Two effectors of PI3K, AKT and glycogen synthase kinase3β (GSK3β), are tightly regulated by ILK (Schlessinger and Hall, 2004). The phosphoinositide-3 kinase-related signaling has indeed been intimately linked to anoikis signaling. ILK interacts with the cytoplasmic tails of β1 and β3 integrin subunits, and is activated transiently by cell-matrix adhesion or growth factor stimulation (Hannigan, Troussard and Dedhar, 2005). The role of ILK in the regulation of AKT Ser473 phosphorylation and activation is complex and is likely to be cell-type and context dependent. Importantly, however, in cancer cells ILK activity is constitutively activated in PTEN-negative human cancers, such as prostate cancer, and the inhibition of ILK activity in these PTEN negative cells impairs downstream AKT targets (Persad *et al.*, 2000). At the clinical level, analysis of ILK expression in human prostate cancer specimens demonstrated that ILK expression levels increased with tumor grade, with an inverse correlation between ILK levels and five-year patient survival (Graff *et al.*, 2001).

Galectins belong to a highly conserved family of animal lectins with members present in organisms ranging from nematodes to mammals (Houzelstein *et al.*, 2004). Galectin overexpression is found in tumor epithelial cells and tumor-associated stromal cells (Lahm *et al.*, 2004; Oka, Takenaka and Raz, 2004), to correlate with acquisition of a metastatic phenotype. The effectiveness of galectin in the malignant transformation of epithelial cells probably resides in a change of its subcellular localization (cytosolic translocation). Cytoplasmic gal-3 is anti-apoptotic, while the nuclear presence of gal-3 exerts pro-apoptotic properties (Oka, Takenaka and Raz, 2004). Opposite functions of gal-3 have been documented in prostate cancer cells depending on their cellular locali-zation: cytoplasmic gal-3 induces cell migration and anchorage-independent cell growth, while nuclear gal-3 exerted the opposite effect, blocking cellular migration (Califice *et al.*, 2004). The dynamics of gal-3 cellular localization are mechanistically intriguing in response to apoptotic stimuli. Translocation from the peri-nuclear membrane to the mitochondria is dependent on synexin (Annexin VII), a phospholipid and $Ca^{++}$ binding protein (Yu *et al.*, 2002). This dynamic sub-cellular localization suggests an additional level of molecular regulation that can exploited for therapeutic targeting. The precise mechanism by which gal-3 regulates cell proliferation is not well understood, but gal-3 can regulate cell-cycle progression by blocking cyclins A and E, and by stimulating p27 and p21 (cell-cycle dependent kinase inhibitors), resulting in cell cycle arrest (Kim *et al.*, 1999). Most importantly, in the context of metastatic spread, is the role defined for galectin as a potent suppressor of anoikis; gal-3 exerts an angiogenic activity by inducing endothelial cell migration (Nangia-Makker *et al.*, 2000) (Fig. 3.3).

TrkB is a neurotrophic tyrosine kinase and a unique receptor that, upon binding to its ligand, brain-derived neurotrophic factor (BDNF), is activated promoting proliferation, differentiation, and survival of retinal and glial cells (Dionne *et al.*, 1998). An early series of studies from the Isaacs laboratory documented that Trk receptor activation by the neurotrophin ligands nerve growth factor (NGF), BDNF, and neurothrophins NT-3 and NT4/5 leads to the initiation of survival mechanisms in prostate tumor cells but not in normal cells (George *et al.*, 1999; Weeraratna *et al.*, 2000). In prostate carcinomas, TrkB and BDNF are frequently over-expressed and there is a close correlation with aggressive behavior and poor prognosis (Aoyama *et al.*, 2001). Elegant genetic studies in rat intestinal epithelial cells identified TrkB as a key suppressor of anoikis, conferring anoikis resistance (Douma *et al.*, 2004). Mechanistic analysis revealed that the TrkB pathway does not activate known downstream targets of AKT such as p70S6 and *Rac*, suggesting that alternative molecules downstream of AKT may perform such a function (Huang and Reichardt, 2003).

Caveolin (cav-1) has been implicated in a variety of cellular processes ranging from signal transduction up to cholesterol homeostasis and lipid transport (Cohen *et al.*, 2004). Caveolins have been directly linked to cancer development and progression (Bender *et al.*, 2000; Razani *et al.*, 2001; Hayashi *et al.*, 2001), thus gaining notoriety as potential predictors of poor prognosis. Cav-1 can indeed act as a tumor-promoter protein in tumors such as bladder, esophageal, and prostate carcinomas (Satoh *et al.*, 2003; Li *et al.*, 2001); up-regulation of cav-1 correlates with increased cell survival, androgen-independence, and increased metastatic potential of prostate tumors (Nasu *et al.*, 1998; Yang *et al.*, 1998; Tahir *et al.*, 2001; Williams *et al.*, 2005). Studies in the transgenic adenocarcinoma mouse prostate (TRAMP) model (Greenberg *et al.*, 1995) confirm that in contrast to mammary tumors, in prostate cancer cav-1 is a tumor promoter, and loss of cav-1 results in apoptosis of prostate epithelial cells via up-regulation of pro-apoptotic genes *PTEN* and Par4 (Williams *et al.*, 2005). Cav-1 is overexpressed in primary tumors, leading to increased cellular proliferation, but the cellular demands are reversed during tumor cell migration as survival requires reduction of cav-1 expression. As schematically illustrated in Fig. 3.3, cav-1 holds a prominent place in the molecular signature of the metastatic phenotype of prostate cancer epithelial cells, a role resonating with evidence that secreted cav-1 indeed contributes functionally to the process of metastasis in castration-resistant prostate cancer (Tahir *et al.*, 2001). Further understanding of the role of cav-1 in metastasis of human prostate cancer cells requires insightful mechanistic consideration of the ability of cav-1 to suppress anoikis by activating the AKT survival pathway, and to also block two potent AKT inhibitors, PP1 and PP2A (Li *et al.*, 2003a). Perhaps the most interesting aspect of cav-1 signaling is the interplay between cav-1 and IGF-1 signaling. Cav-1 is a known mediator of the insulin and IGF-1 signaling pathway, and enhances matrix-independent cell survival by upregulating IGF-1R (a mitogenic anti-apoptotic receptor) expression, and signaling (Baserga, Peruzzi and Reiss, 2003). Increased IGF-1 signaling promotes cell survival via activation of the AKT survival pathway, while targeting cav-1 expression in prostate cancer cells, and significantly enhances androgen deprivation-induced apoptosis (Li *et al.*, 2003a).

The oncogenic effect of IGF resides mainly in its function as a potent inhibitor of apoptosis, by blocking the induction of apoptosis by TGF-β, and by activation of AKT (Song *et al.*, 2003). Elevated levels of IGF-1 have been correlated to prostate cancer development in transgenic mice, while enforced overexpression of IGF-1 promotes prostate carcinogenesis (Di Giovanni *et al.*, 2000). Furthermore, the IGF signaling pathway has been implicated in conferring resistance to anoikis in a variety of cancer cells (Dupont *et al.*, 2001; Ravid *et al.*, 2005; Goel *et al.*, 2005). There is also evidence to link IGF-1-mediated resistance mechanistically to anti-androgen therapy by up-regulating survivin, an IAP apoptosis family member directly involved in cancer progression and drug resistance (Altieri, 2003). One must pause to consider the consequences of high expression IGF-1 and IGF-1R in the bone in terms of this signaling pathway playing a vital role in the successful seeding of metastatic prostate cancer cells to the bone, as well as osteoblast proliferation (Yoneda and Hiraga, 2005). Unlike other tumors that metastasize to the bone (for instance those affecting the breast and lung), prostate cancer causes mostly osteoblastic instead of osteolytic lesions (Logothetis and Lin, 2005). Therefore, targeting the IGF-1 signaling in prostate cancer by affecting tumor cell resistance to anoikis and metastatic tumor cell deposits to the bone provides unique and most significant therapeutic possibilities which call for full exploitation.

The intersection between mechanotransduction pathways with oncogenic signaling pathways is receiving increased recognition as an important synergy in promoting tumor initiation and progression to metastasis. Mechanical properties of the microenvironment are sensed by integrin family receptors that connect ECM proteins outside the cell to the actin cytoskeleton inside the cell (Chen, 2008). Increased tension from a stiffer matrix induces integrin clustering, development of focal adhesions, and activation of multiple downstream signaling pathways. The focal adhesion complex contains an interactive matrix of numerous proteins including non-receptor tyrosine kinases, such as Src and focal adhesion kinase (FAK), and adaptor and actin binding proteins including talin and paxillin, as well as cytosolic phosphatases and proteases. Specifically, the calpain proteases have been implicated in the cleavage of focal adhesion proteins that promote focal adhesion turnover (Dourbin *et al.*, 2001; Hauck *et al.*, 2001). FAK phosphorylation is known to signal the Rho pathway, which influences cell contractility by mediating cytoskeletal processes such as focal adhesion and actin stress fiber formations (Chen, 2008). More recent evidence indicates that a close interaction between the focal adhesion complex and caspase-8 (normally an apoptosis initiator caspase) promotes tumor cell migration and metastasis (Barbero *et al.*, 2009).

Several proteins critical to the regulation of cell survival, including the focal adhesion complex such as FAK, and intracellular effectors such as Ras/ERK, are activated by integrin adhesion, and detachment from the ECM via anoikis results in their inactivation. Thus enhanced FAK activation on a stiffer matrix can also increase phosphorylation of extracellular signal-regulated kinase (ERK) via Ras activation (Provenzano *et al.*, 2009). ERK can control cell migration, invasion, proliferation, and differentiation through modulation of acto-myosin contraction as well as induction of transcriptional programs, differentially responding to the stiffness of the microenvironment (Provenzano *et al.*, 2009). Talin — the newest player of the focal adhesion complex being recognized as an anoikis resistance regulator — merits special functional interrogation. High talin1 expression enhances the migration and invasion potential of human prostate cancer cells by engaging an AKT-signaling activation, while loss of talin1 reduces the invasion potential and *in vivo* metastatic ability of prostate cancer cells. Significantly enough, talin1 mRNA is downregulated by androgens in primary prostate cancer cells (Sakamoto and Kyprianou, 2010). Taken together the evidence supports an indirect link between talin1 and prostate cancer metastasis and emergence to androgen independence. The unfolding mechanistic scenario is intriguing. Overexpressing talin1 in prostate cancer cells leads to activation of FAK/AKT signaling through both ECM-dependent and ECM-independent mechanisms. Activation of AKT survival signaling has been causally linked to anoikis resistance in detached cells (Farhana *et al.*, 2004). In direct accordance with this concept, we recently established talin1 as a promoter of tumor cell metastasis by conferring anoikis resistance, invasive properties, and survival after detachment from the primary tumor site through engaging the FAK and AKT signaling (Sakamoto *et al.*, 2010). This will not be the first case of key focal adhesion protein being assigned a role as a cancer metastasis enhancer: paxillin overexpression is shown to correlate with meta-static and invasion properties in hepatocellular carcinoma (Li, Shi and Zhang, 2005). In a similar mechanistic connection, recruitment of FAK and paxillin to β1 integrin promotes cancer cell migration through MAPK activation (Crowe and Ohannessian, 2004). The rapidly growing evidence indicates that talin1 is causally involved not in primary tumor development, but rather in promoting local invasion, and ultimately distant metastasis, by conferring resistance to anoikis (Fig. 5.1). Resistance to anoikis, possibly conferred by talin1, is not a selective advantage during tumor growth in an ECM-rich microenvironment; rather during the invasion through vasculature during metastasis, resistance to anoikis could predict a greater ability to form metastatic lesions. A similar functional contribution in enhancing metastasis but not primary tumor growth is shown by the apoptotic suppressor of cytoskeleton-dependent death, BCL-XL, during mammary tumorigenesis (Martin *et al.*, 2004).

The primary target of interference by talin1 is the binding of integrin-linked kinase (ILK-1) to the integrin β3 subunit, a component of the focal adhesion complex engaged in promoting angiogenesis, invasion of surrounding tissues, and metastasis. ILK is a serine/threonine protein kinase that interacts with the

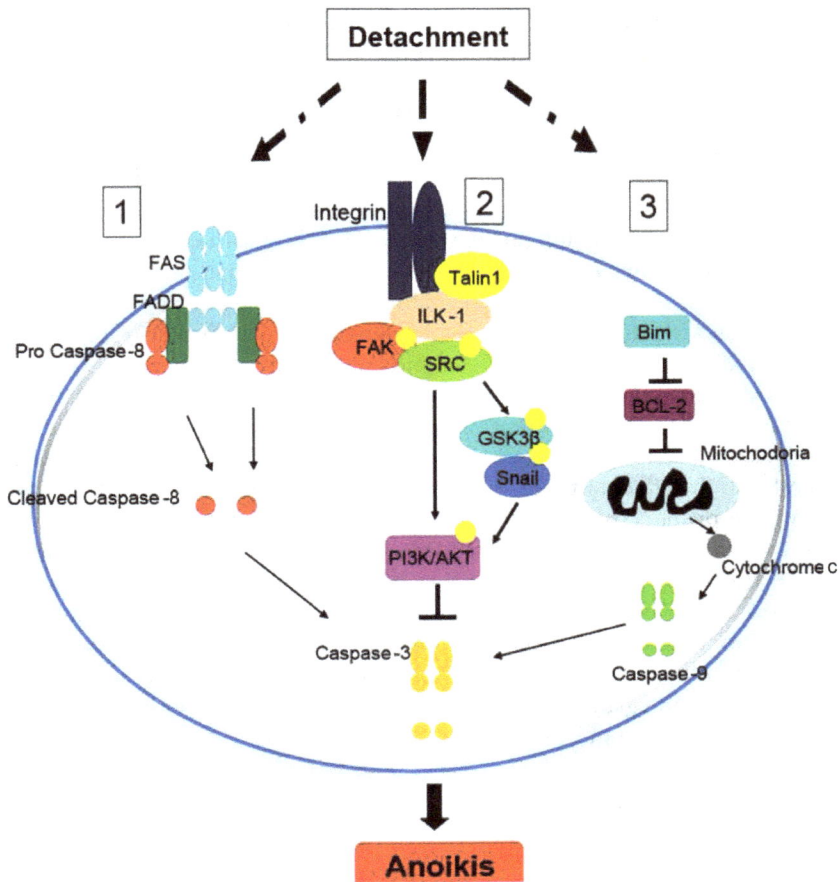

**Figure 5.1** Programming of the anoikis phenomenon. Upon cell detachment, anoikis signaling can be activated via three major pathways: (1) Death receptor mediated pathway; (2) ECM-integrin cell survival pathway; and (3) Mitochondrial mediated pathway. (1) Activation of FAS/FADD in death receptor induces cleavage of caspase-8 and subsequently activates downstream caspase-3 to induce apoptosis. (2) ECM-epithelial cell interaction mediates integrin-dependent activation of talin1, FAK/SRC and ILK-1. Activation of ILK-1 mediates phosphorylation of GSK3β and recruitment of *Snail*. This signaling cascade activates PI3K/AKT cell survival signals. Loss of survival signals upon cell detachment enhances cell susceptibility to apoptotic stimuli. (3) Cell detachment induces up-regulation of pro-apoptotic proteins, such as Bim, and down-regulation of anti-apoptotic regulatory proteins such as BCL-2, which activate the cytochrome c release from the mitochondria and subsequently trigger activation of caspase-9 and caspase-3 towards apoptotic execution of the cell.

cytoplasmic domain of β1-integrin and β3-integrin, and has been functionally linked to integrin and Wnt signaling pathways (Hannigan *et al.*, 1996; Wu *et al.*, 2001). ILK regulates several integrin-mediated cellular processes including cell adhesion, fibronectin-ECM assembly, and anchorage-dependent cell growth (Lynch *et al.*, 1999; Fukuda *et al.*, 2003; Hannigan, Troussard and Dedhar, 2005). Upon cell adhesion, ILK is transiently activated and directly phosphorylates AKT Ser473 and glycogen synthase kinase-3 (GSK3β) (Atwell, Roskelley and Dedhar, 2000); in contrast, inhibiting ILK in cancer cells inhibits AKT phosphorylation and cell survival (Fig. 5.1). Additional evidence reinforces the critical role of ILK in anoikis regulation; ILK-1 mediates anoikis resistance even without activation of integrin/ECM signals, possibly by recruiting through parvin-mediated targeting of AKT to lipid rafts (Fukuda *et al.*, 2003; Hannigan, Troussard and Dedhar, 2005). Recent *in vitro* studies indicate that talin1 confers anoikis resistance (non-adherent) and mediates direct ECM-epithelial interaction (adherent conditions) (Sakamoto *et al.*, 2010). Taken together, the data suggest that the effect of talin1 on anoikis might be responsible for the acquisition of the primary tumor cell invasive and metastatic properties leading to prostate cancer metastasis. As illustrated in Fig. 5.1, talin1 binding to β integrin recruits the focal adhesion proteins ILK, FAK and SRC, sealing its link with the ECM towards activation of the downstream signals including AKT and ERK. This signaling activation promotes cell survival, invasion, and angiogenesis. Under non-adherent conditions, talin1 stimulates FAK, SRC, and GSK3-β independent of integrin signaling, and confers resistance to anoikis, leading to metastatic spread of cancer cells.

## 5.6 Significance of Apoptosis in Cytoskeleton and Microtubule Targeting

Alterations in cytoskeleton reorganization induced by diverse signals (including androgens and growth factors such as TGF-β) may enable cell migration and metastasis of escaped prostate tumor cells by impacting apoptosis outcomes. Changes in actin microfilament network organization in androgen-treated cells could provide active movement assisting cell migration and the dynamics of interaction with adherent molecules in the ECM (Zhu *et al.*, 2010b). Cofilin is a novel player identified, by functional proteomic analysis, as an effector of TGF-β apoptotic signaling in prostate cancer cells (Zhu *et al.*, 2006); it is an actin-binding protein that promotes actin polymerization, directs cell migration, and, via its mitochondrial translocation, represents an early event in apoptosis induction (Ono, 2003). The apoptosis-inducing ability of cofilin, but not its mitochondrial localization, is dependent on the functional actin-binding domain, supporting the concept that the domains involved in the mitochondrial targeting and actin binding are indispensable for its pro-apoptotic function.

Consistent with this concept is evidence indicating cofilin translocation from the cytosol to the mitochondria in response to TGF-β, an event preceding cytochrome c release from the mitochondria (Zhu *et al.*, 2006). Cofilin as the major calcium-independent regulator of the dynamics of actin cytoskeleton assembly/disassembly plays a prominent role, not only in apoptosis induction via its targeting to mitochondrial membranes in response to apoptotic stimuli (Chua *et al.*, 2003), but also in cell movement and migration (Ono, 2003; DesMarais *et al.*, 2005), and reorganization of actin cytoskeleton in response to diverse extracellular stimuli (Wang, Shibasaki and Mizuno, 2005).

The tubulin/microtubule system, an integral component of the cytoskeleton, is a therapeutic target for prostate cancer treatment (Mancuso, Oudard and Sternberg, 2007). Microtubules are highly dynamic structures that play a critical role in orchestrating the separation and segregation of chromosomes during mitosis (Giannakakou, Sackett and Fojo, 2000; Hammond, Cai and Verhey, 2008). Once the motor protein Kinesin-1 is recruited to the microtubules, it preferentially moves various cargoes, including vimentin filaments and transferin, along detyrosinated microtubules (Liao and Gundersen, 1998; Kreitzer, Liao and Gundersen, 1999). Tubulin-binding agents are derived from natural sources and include a large number of compounds with diverse chemical structures, all sharing an ability to disrupt microtubule dynamics, induce mitotic arrest, and promote apoptosis (Giannakakou *et al.*, 2000). The best characterized of these agents are the *vinca* alkaloids and taxanes, which at high doses cause microtubule destabilization and microtubule stabilization, respectively. Phase I and phase II clinical trials established taxane-based chemotherapy as the only treatment currently approved for patients with CRPC, but with a relatively modest survival benefit (Berry *et al.*, 2001; Petrylak *et al.*, 2004; Beer *et al.*, 2004).

Two independent multicenter phase III studies [Southwest Oncology Group (SWOG) 99-16 and TAX 327] compared taxane-based regimens with Mitoxantrone/Prednisone and demonstrated a significant survival benefit in patients with CRPC (Petrylak *et al.*, 2004; Tannock *et al.*, 2004). Docetaxel, a semisynthetic taxane, stabilizes the microtubule by promoting binding to β-actin. Once bound, microtubules cannot be disassembled, thereby disrupting mitosis, causing G2M cell cycle arrest, and triggering apoptosis (Kraus *et al.*, 2003). Both Docetaxel and prednisone chemotherapy have become first-line standard treatments of metastatic CRPC (Beer *et al.*, 2001). The potentially enhanced efficacy of these therapeutic agents in combination with other chemotherapeutic agents targeting advanced CRPC is still the focus of several clinical investigations (Beer *et al.*, 2001; Petrylak, 2005; Mancusco, Oudard and Sternberg, 2007).

The long-term benefit of androgen deprivation in patients with localized and metastatic disease, and the recent breakthroughs in the development of novel

AR-antagonist strategies with the potential to improve the therapeutic outcome and survival in patients with CRPC, have been discussed above (Lu-Yao *et al.*, 2008; Tran *et al.*, 2009). The clinical knowledge of Paclitaxel as the only effective treatment for CRPC emphasizes the need to understand the mechanisms of action of this drug in order to augment its therapeutic efficacy. Considering the requirement of AR signaling to drive prostate growth and survival, and since CRPC still retains AR activity, work by our group explored the impact of tubulin and microtubule-targeting drugs on AR signaling in prostate cancer.

Microtubule stabilization through binding of Docetaxel to $\beta$-tubulin is the traditionally accepted mechanism of action. Once bound with taxanes, micro-tubules cannot disassemble, thus the static polymerization disrupts the normal mitotic process, arrests cells in G2M phase, ultimately inducing apoptotic cell death. Of apparent clinical relevance is the discovery that taxanes can target prostate tumors via alternative routes besides mitosis disruption. Notably Docetaxel counteracts the pro-survival effects of BCL-2 gene expression (Debes and Tindall; 2004; Oliver *et al.*, 2005). Thus treatment of prostate cancer cells overexpressing BCL-2 with taxol induces BCL-2 phosphorylation. BCL-2 phos-phorylation inhibits its binding to BAX and consequently apoptosis of prostate cancer cells in response to taxol (Haldar *et al.*, 1996; Basu and Haldar, 1998). The recent study by Zhu and colleagues indicates yet another intriguing mechanism triggered by taxane-based regimens, towards impairing nuclear localization and activity of AR. The findings establish that microtubule-targeting chemotherapy drugs could inhibit both androgen-dependent and androgen-independent activation of AR, by blocking AR nuclear translocation (Zhu and Kyprianou, 2010). Thus, it is likely that addition of an AR-binding moiety to a therapeutic agent such as Taxol could selectively target AR-expressing prostate cancer cells. A potential combination of ADT and taxanes may augment efficacy by targeting both androgen-dependent and androgen-independent prostate tumor growth.

Modification of tubulin (detyrosination/tyrosination) can affect the microtubule stability. A feedback loop may explain the reduced association and co-localization between AR and tubulin, due to tubulin down-regulation. Androgen signaling is important in cell differentiation, and regulates the cell cycle including G2M arrest (Heisler *et al.*, 1997; Litvinov *et al.*, 2006), consistent with its function in inhibiting microtubule structures. In addition, molecular dynamics-based studies indicated that a conjugate of colchicine and an AR antagonist (cyanonilutamide) with tubulin-inhibiting activity increases cytoplasmic AR levels, and anatagonizes AR activity in prostate cancer cells (Sharifi *et al.*, 2007). Moreover, indirect support for a microtubulin-targeting action influencing steroid receptor activity is gained from evidence on the ability of estrogens to regulate $\beta$-tubulin synthesis and decrease microtubule density, ultimately blocking prostate cancer cells at G2M phase (Bonham *et al.*, 2002; Montgomery *et al.*, 2005). The recent findings by Zhu and colleagues (Zhu and Kyprianou, 2010) clearly establish that microtubule-targeting agents play a prominent role in impairing AR nuclear transport and activity, thus promoting prostate tumor suppression. This evidence is of strong translational significance, as it hints to a potential new mechanism underlying treatment failure of prostate cancer patients to Paclitaxel within the microtubule repertoire in CRPC.

## 5.7 Autophagy: The Cellular Benefits of Starving to Death

Dissection of critical anoikis signaling events enables anoikis and inhibits cell survival during the state of "homelessness", thus effectively impairing invasion and metastasis. However, there is a cautionary caveat as novel regimens purporting to restore anoikis must consider the exchanges between the two main mechanisms of anoikis-related cell death — apoptosis and autophagy. As discussed in the introductory chapter, apoptosis has only one outcome — the orderly elimination of the cell via activation of its suicidal molecular programs. In marked contrast, the process of autophagy is much more functionally diverse in its effects, serving as both a cell survival and a cell death mechanism depending on the context, the stimuli and the extent of autophagy, which are likely exploitable for cancer research. For anoikis to be thwarted, and cancer cells to succeed in their metastatic journey, tumors must stumble upon the right way to both arrest apoptosis and use autophagy in a way that promotes their survival (Horbinski, Mojesky and Kyprianou, 2010). Thus, understanding the functional cross-talk between apoptosis and autophagy signaling pathways is a prerequisite for effective targeting of metastasis.

Autophagy (from the Greek, to "eat oneself") refers to any cellular degradative pathway that delivers cytoplasmic cargo to the lysosome (Levine and Klionsky, 2004). The newness of autophagy in cancer research is underscored by the fact that, in PubMed, nearly 70,000 cancer-oriented papers to date deal with apoptosis while only about 1,000 deal with autophagy. Autophagy is triggered by nutrient deprivation, starvation, and other stress stimuli, upon which the cell breaks down and recycles vital internal components (Eisenberger-Lerner *et al.*, 2009). This process of self-digestion by the cells not only provides nutrients to maintain vital cellular functions during fasting, but it can also rid the cell of superfluous or damaged organelles, misfolded proteins and invading microorganisms (Levine and Kroemer, 2008). Unlike the ubiquitin–proteasome system, which degrades short-lived proteins, macroautophagy (referred to as autophagy) processes long-standing macromolecules and whole organelles, most notably the mitochondrion (aka mitophagy). This highly conserved process of autophagy is divided into four phases. The first phase, called induction, occurs in response to environmental stressors like nutrient deprivation, oxidative stress, infection, or hypoxia. This leads to the second phase, wherein structures targeted for degradation are enveloped by a membrane called an autophagophore. The resulting double-layered membrane is an autophagosome. Third, this autophagosome fuses with a lysosome containing hydrolytic enzymes, forming an autophago-lysosome. The fourth and final phase consists of the actual enzymatic degradation and recycling of materials (Horbinski, Mojesky and Kyprianou, 2010). Thus the intracellular digestion of macromolecules during the autophagic process is a self-conserved response.

The dynamics of occurrence of the autophagic phenomenon, although not completely understood, are obviously highly sophisticated and tightly regulated, involving a number of complex-forming proteins (Levine and Kroemer, 2008). Under pro-autophagic conditions like amino acid deprivation, Beclin 1 (ATG6 in yeast cells), PI3K III, and Vps34 aggregate and promote induction. A group of autophagy (ATG) proteins, including ATG12, ATG5, and ATG16, form a complex via ATG7 and ATG10, the latter two proteins acting in a manner analogous to ubiquitin-like enzymes E1 and E2, respectively. Next, microtubule-associated protein 1 light chain 3 (LC3/ATG8) is activated by ATG4 and ATG7. LC3 is then ligated to phosphatidylethanolamine (PE) via ATG3. The ATG12-ATG5-ATG16 complex, in conjunction with LC3-PE, facilitates autophagophore membrane elongation and autophagosome formation (Eisenberger-Lerner *et al.*, 2009). A key regulator of autophagy is the mammalian target of rapamycin (mTOR) which, upon activation, becomes a potent inhibitor of ATG12-ATG5-ATG16 complex formation (Pattingre *et al.*, 2008).

The knowledge gathered so far from molecular dissection of apoptosis and autophagy signaling would allow for combination targeting of the converging points in apoptosis, autophagy, and anoikis death modes. The cross-talk between anoikis and apoptosis converges at the key activation point of CD95/FAS as it connects RIP, a kinase that shuttles between CD95/FAS-mediated death and FAK-mediated survival pathways (Horbinski, Mojesky and Kyprianou, 2010). The functional exchange coordinator, Beclin 1, is the key point of the cross-talk between apoptosis and autophagy. This protein, known as ATG6 in yeast, comprises an early component of the autophagic vesicle that is destined to later become an autophagosome. Beclin 1 contains a BH3 (BCL-2 homology region-3) domain capable of binding to the antiapoptotic BCL-2/BCL-XL proteins (Mailleux *et al.*, 2007). When this happens, ATG6/Beclin 1 activity and subsequent autophagosome formation can be blocked (Pattingre *et al.*, 2008). On the other hand, the proapoptotic BH3-only domain protein BAD directly removes this inhibition, thereby facilitating both autophagy and apoptosis (Mailleux *et al.*, 2007). Adding to the complexity of these interactions is the finding that BAX, a proapoptotic BH1-3 protein, inhibits autophagy by promoting caspase-dependent cleavage of Beclin 1, an interaction that is itself completely prevented by BCL-XL (Zong *et al.*, 2001). The intriguing mechanistic twist is that autophagy stimulated by ECM detachment is not sensitive to BCL-2 (Fung *et al.*, 2008), which suggests that in anoikis there is variable sensitivity to such cross-talk. The subcellular distribution/localization of BCL-2 is also a critical determining factor, as it blocks autophagy when localized to the endoplasmic reticulum, and blocks apoptosis when associated with the mitochondria (Tomita *et al.*, 2006). Mechanistically, activated ERK1/2 also triggers autophagy of damaged mitochondria, and may even exert signaling effects from within the organelle (Yu *et al.*, 2001). As with apoptosis, the net effect of ERK in autophagy depends on the activity and temporal variations leading to cell death rather than cell survival (Matthew *et al.*, 2009). The emerging scenario is intriguing: depending on its level and the timing of its activation, ERK can either promote or antagonize autophagy and/or apoptosis, likely serving as a pro- versus anti-anoikis switch.

DAPk is a serine-threonine cytoskeleton-associated kinase that sensitizes the cell to a variety of apoptosis-inducing signals, including activation of p53 and sequestration of ERK, in addition to blocking integrin/FAK-dependent cell survival (Chen *et al.*, 2005). This protein also promotes lethal autophagy in response to cellular stresses, at least in part by directly phosphorylating Beclin 1, freeing it from BCL-XL (Zalckvar *et al.*, 2009a). DAPk however can also bind to and disrupt the TSC1/TSC2 complex, resulting in mTOR activation and reduction in autophagy upon exposure of HEK293 cells to certain growth factors (Zalckvar *et al.*, 2009b). Thus, while autophagy can facilitate apoptosis by generating enough ATP to fuel membrane blebbing, it may also prevent apoptosis by destroying damaged mitochondria, thereby limiting ROS exposure and cytochrome c release (Yu *et al.*, 2001). Impaired autophagy could promote tumorigenic development by allowing accumulation of mutation-inducing byproducts, insofar as Beclin 1 haploinsufficiency has been identified in many carcinomas, and transgenic mice lacking one copy of Beclin 1 show marked neoplastic diathesis (Yue *et al.*, 2003). At the molecular level it has been established that inhibition of Beclin 1 and autophagy causes accumulation of p62, elevated ROS, and increased chromosomal instability, resulting in tumorigenesis, including prostate cancer (Matthew *et al.*, 2009; Moscat and Diaz-Meco, 2009).

The functional interplay platform is richly populated by the primers of death, and detailed dissection of their role (antagonistic or synergistic) has potentially crucial consequences for the development of novel anti-tumor therapies aimed at restoring anoikis and inhibiting metastasis. Autophagy antagonizes the proapoptotic effects of melanoma differentiation-associated gene-7/interleukin-24 gene therapy in leukemia cells. In colon cancer cells, inhibiting autophagy with chloroquine allows the histone deacetylase inhibitor vorinostat to induce apoptosis (Carew *et al.*, 2010), while trifluorothymidine appears to trigger apoptosis more effectively than 5-fluorouracil as it elicits less autophagy (Bijnsdorp *et al.*, 2010). Glioma stem-like, CD133-positive cells are more resistant to radiation-induced apoptosis, at least in part because of increased autophagy (Lomonaco *et al.*, 2009).

Beclin 1, ROS and ERK are the critical molecules responsible for integrating autophagy and apoptosis, and whose effects can differ greatly from one situation to another, challenging selective therapeutic targeting (Apel *et al.*, 2009). The functional involvement of autophagy in ATP-dependent generation of engulfment signals and heterophagic removal of apoptotic corpses hints at a potentially important role for autophagy in preventing inflammation (Qu *et al.*, 2007). The rapid removal of apoptotic corpses is critical for prevention of tissue inflammation and defective autophagy may contribute to inflammation-driven prostate tumorigenesis. Activation of autophagy, however, is not without risks as it may keep alive cells that should die, such as chemotherapy-treated tumor cells. Indeed, recent experimental studies documented that autophagy inhibition effectively enhanced cell death induced by SRC family kinase (SFK) inhibitors (Wu *et al.*, 2010). Moreover, it was demonstrated in the same study that a combination of saracatinib with chloroquine significantly reduced prostate cancer growth *in vivo*, which is strong evidence autophagy serves a protective role in SFK inhibitor-mediated apoptosis (Wu *et al.*, 2010). Therefore, clinically acceptable autophagy modulators may be used as adjunctive therapeutic agents for SFK inhibitors, with potentially enhanced therapeutic benefit in prostate cancer patients. This is of high clinical impact as SRC activity is associated with the development of CRPC.

Mechanistically exciting new evidence recently reported by Richard Pestell's group suggests that the "longevity" protein SIRT1, known for its life-spanning effects in different species, acts as a checkpoint of autophagy towards impairing prostate tumor growth by inhibiting the development of premalignant precursor (PIN), pathologically associated with reduced autophagy (Powell *et al.*, 2011). SIRT1 induction of autophagy occurring at the level of autophagosome maturation was demonstrated *in vivo* and *in vitro*. Thus, elucidation of the most effective method to manipulate autophagy for prostate cancer therapy will require careful investigation *in vivo*. Particular focus on targeting cellular metabolic pathways may provide a multifaceted approach for the effective impairment of cancer growth and metastasis.

# 6

# Epithelial–Mesenchymal Transition (EMT) in Prostate Cancer Metastasis

Understanding the metastatic process requires detailed understanding not only of tumor epithelial cell behavior and growth characteristics, but also the communication between cells and their neighbors in the adjacent environment (Hart *et al.*, 2005). Such global communications by the epithelial cell layer is maintained through gap junctional complexes separated from adjacent tissue by a basal lamina. As discussed in Chapter 2, mesenchymal cells are loosely organized in a three-dimensional ECM comprising connective tissue adjacent to the stroma. During tumor progression to metastasis, profound phenotypic changes affect these structural dynamics of tissue plasticity: the epithelial–mesenchymal transition (EMT) (Janda *et al.*, 2002; Hugo *et al.*, 2007), and the formation of a reactive stroma, reflecting intense rearrangement of the ECM in the spatial confines of the tumor cell cytoskeleton (Chung *et al.*, 1989). EMT is a coordinated molecular and cellular change defined as a profound reduction in cell–cell adhesion, apical-basolateral polarity, and epithelial markers; and a simultaneous acquisition of cell motility, spindle-cell shape, and mesenchymal markers. During EMT, polarized epithelial cells acquire a migratory fibroblastoid phenotype and that is universally encountered during both physiological and pathological conditions (Thiery, 2002); EMT also establishes its significance as a critical event during cancer metastasis (Thompson, Newgreen and Tarin, 2005). Such transitions, necessary for proper embryonic development, provide a convenient venue for epithelium-derived tumors to become highly invasive and metastasize rapidly (Thiery, 2003; Tarin, Thompson and Newgreen, 2005), following a mechanism resembling a reawakening/reactivation of the embryonic program of EMT. The most important similarity between embryonic and tumorigenic EMT is that in both cases, the migrating cells "re-evaluate" their surroundings, and change their structural relationship with the ECM. EMT endows cells with migratory and invasive properties, induces stem cell properties and prevents apoptosis and senescence, thus subtly orchestrating the initiation events towards exploitation of the microenvironment (Christophori, 2003) and manifestation of metastasis (Tarin, Thompson and Newgreen, 2005; Thompson, Newgreen and Tarin, 2005). Even if there is no histological evidence of cellular intermediates during the transition from epithelial to mesenchymal phenotype (Tarin, Thompson and Newgreen, 2005), the key factor is that activation of mesenchymal genes in epithelial cells is crucial for EMT success (Thompson, Newgreen and Tarin, 2005; Barrallo-Gimeno and Nieto, 2005). The hallmark of EMT is loss of expression of the cell adhesion molecule E-cadherin and epithelial cell marker $\beta$-catenin, and gain of mesenchymal-cell markers N-cadherin and vimentin at the leading edge of the invasive front of solid tumors. E-cadherin is a cell–cell adhesion molecule that participates in calcium-dependent interactions to form epithelial adherent junctions (Aclogue *et al.*, 2009). To preserve cellular shape and polarity the intracellular domains of cadherins connect to the actin cytoskeleton through $\alpha$-catenin and $\beta$-catenin. It is the signaling activities of the mesenchymal cells that facilitate migration and survival in an anchorage-independent environment.

Prostate epithelial cells undergo EMT in response to an array of soluble factors including TGF-$\beta$1 plus EGF, IGF-1, $\beta$2-microglobulin ($\beta$2-m), or exposure to a bone microenvironment (Zhau *et al.*, 2008). Cadhein switching during prostate cancer progression becomes a fascinating scenario with distinct functional consequences on tumor cell behavior (Tomita *et al.*, 2000). In an intriguing twist it was recently documented that androgens suppress E-cadherin expression, while inducing the mesenchymal marker expression in prostate cancer epithelial cells (Cavallaro and Christophori, 2004; Zhou and Kyprianou, 2010). This might facilitate escape of prostate cancer cells from the primary site and migration to distant sites, an important concept because activation of EMT may result in increased bone turnover, implicated in prostate cancer bone colonization during metastasis. Considering that the reactive prostate stroma has been assigned a critical role in the context of the tumor microenvironment in prostate cancer progression to metastasis, AR signaling in prostate fibroblasts may function as a promoter of prostate epithelial cell proliferation (Niu *et al.*, 2001), as well as a mediator of a functional exchange between prostate epithelial and stromal cells, thus contributing to the EMT effect during cancer metastasis (Tlsty and Hein, 2001; Zhu and Kyprianou, 2008).

The existence of rigorous and functionally productive cross-talk between the androgen axis and TGF-$\beta$ signaling has been firmly established (Djakiew, 2000; Zhu and Kyprianou, 2008). In human prostate cancer cells PC-3 and LNCaP (Horoszewicz *et al.*, 1983), Smad3 enhances AR transactivation, while co-transfection of Smad3 and Smad4 repress AR transactivation (Bruckheimer and Kyprianou, 2001; Kang *et al.*, 2002; Zhu *et al.*, 2008). The interaction between the androgen axis and TGF-β signaling could be the determining factor for EMT manifestation (Salm *et al.*, 2000; Schiemann *et al.*, 2002). Nuclear translocation of β-catenin has been reported in the invasive front of colorectal carcinoma (Brabletz *et al.*, 2001). Moreover, β-catenin activates DNA binding protein LEF-1/TCFs to induce several signaling pathways towards mesenchymal marker expression (Eger *et al.*, 2000). A functional exchange between AR and β-catenin results in increased nuclear colocalization, and interaction of AR with β-catenin, in castration-resistant prostate tumors (Cronauer *et al.*, 2005; Wang, Wang and Sadar, 2008). Activation of β-catenin by androgen signaling could be an alternative mechanism of androgen-induced EMT in prostate tumor epithelial cells. The involvement of several transcriptional factors (such as zinc-finger factors *Snail* and *Slug*, two handed zinc-finger factors ZEB1 and SIP1, and basic helix-loop-helix factors *Twist* and E12/E47) in the EMT process, by repressing E-cadherin expression and consequently inducing migration and metastasis, has been established (Barallo-Gimeno and Nieto, 2005; Peinado, Olmeda and Cano, 2007; Horiguchi *et al.*, 2009). Downstream activation of *Snail* by TGF-β/Smad pathway represses E-cadherin expression in several cancer cell types (Nieto 2002; Thuault *et al.*, 2008), while *Twist* regulates cytokine gene expression via repression of NF-κB activity (Sosic *et al.*, 2003). In androgen sensitive TGF-β-responsive prostate cancer LNCaP TβRII cells, androgens can independently induce EMT, bypassing the effect by TGF-β (Zhu and Kyprianou, 2010).

The functional outcome of EMT in prostate cancer progression is likely to be complex, given the uncertainty surrounding the contribution of the androgen axis to prostate cancer metastasis. The impact of androgen suppression to metastatic dissemination of prostate cancer cells has been debated as a controversial issue. Thus one could speculate that a low threshold AR level may promote EMT, ultimately facilitating metastatic spread of prostate tumor epithelial cells. The inhibition of EMT response to androgens by AR overexpression points to:

a) an inverse relationship between AR content and EMT induction; and
b) a potential biochemical basis for the metastatic behavior of prostate cancer cells from recurrent androgen-resistant tumors.

Long-term androgen deprivation may downregulate AR expression and facilitate EMT, thus promoting cancer metastasis, and this resonates with the evidence that intermittent androgen deprivation therapy benefits patients during prostate cancer progression (Boccon-Gibbod *et al.*, 2007). Ongoing clinical trials point to intermittent androgen deprivation therapy as a promising option for patients with metastatic prostate cancer, in accord with pre-clinical evidence that intermittent androgen deprivation delays androgen independence (Bruchovsky *et al.*, 2001; Suzuki *et al.*, 2008). Gain-of-function studies indicate that activated

AR (via mutational activation or ligand independent activation) promotes proliferation of prostate cancer cells (Burnstein, 2005; Balk and Knudsen, 2008; Knudsen and Scher, 2009). In a "double-edged sword" twist, loss of AR may actually promote tumor cell metastatic ability by regulating EMT; thus one could argue that the androgen-mediated EMT effect, as a biological process, contributes significantly to prostate cancer metastasis.

TGF-β is a primary effector of EMT during development and during tumorigenesis (Akhurst and Derynck, 2001). This functionally "promiscuous" cytokine was the first EMT inducer described in normal mammary epithelial cells by signaling through its receptor serine-threonine kinase complex; interestingly it remains the best-characterized inducer of EMT in numerous physiological and pathological conditions (Nieto, 2002). Other molecules such as *Snail*, *Twist*, Par6 and NF-κB coordinate with TGF-β to regulate EMT (Xie *et al.*, 2004; Thiery and Sleeman, 2006). Indirect evidence that EMT correlates with poor prognosis in prostate cancer subjects includes loss or delocalization of junctional E-cadherin, replacement of E-cadherin by N-cadherin, degradation of cell–cell adhesion (Tomita *et al.*, 2000), structural polarity and tissue architecture, *Snail* or *Slug* expression, and expression of mesenchymal markers such as vimentin (Thuault *et al.*, 2008). The tumor promotion action of TGF-β during prostate tumorigenesis is functionally consistent with its EMT-inducing activities and plays a critically important role in tumor progression to metastasis. Upon binding by the ligand TGF-β, the TβRII receptor, interacts with claudin, a component of the tight junction and phosphorylates Par6 protein. This direct protein–protein interaction and subsequent Par6 phosphorylation recruits *Smurf1*, thereby leading to ubiquitin-dependent degradation of RhoA, a GTPase family member responsible for the maintenance of apical–basal polarity and junctional stability (Ozdamar *et al.*, 2005). TGF-β signaling also induces *Slug* expression, which inhibits gene expression of desmoplakin and plakoplagin, the two desmosomal plague proteins. This inhibition ultimately leads to disassembly of desmosomes, a specialized type of junction complex functionally responsible for cell-to-cell adhesions (Thiery, 2002).

*Snail* belongs to a zinc-finger family of transcriptional factors that are essential in embryonic development, more specifically during EMT of mesoderm specification (Nieto, 2002). *Snail* binds to E-cadherin promoter and represses its transcription during EMT. Loss of E-cadherin is a critical step towards malignant development and tumor progression in prostate cancer (Schlessinger and Hall, 2004). *Snail* activates the transcription of genes that are associated with mesenchymal differentiation, such as vimentin and fribronectin (Nieto, 2002). Furthermore, during EMT, *Snail* is responsible for suppressing cell proliferation by blocking cell cycle progression and conferring apoptosis resistance, by activating survival genes such as the PI3/AKT cascade, and by inhibiting caspase-3 activation through TNF-α (Barrallo-Gimeno and Nieto,

2005; Vega *et al.*, 2004). Elevated *Snail* mRNA expression has been detected in a variety of tumors; however the level of *Snail* protein is only modestly up-regulated in these tumors. *Snail* is phosphorylated by GSK-3β kinase and this phosphorylation is responsible for regulating the rate of *Snail* degradation via the proteosome (Zhou *et al.*, 2004; Schlessinger and Hall, 2004). Therefore, during EMT, GSK-3β needs to be inhibited so *Snail* can freely perform its functions. Indeed, signaling through EGF, IGF, and hepatocyte growth factor (HGF) causes phosphorylation of GSK-3β, allowing nuclear translocation of *Snail* and subsequently activating its transcriptional repressor/activator function and promoting EMT (Schlessinger and Hall, 2004). Therefore it might not be coincidental that growth factor signaling pathways regulating GSK-3β activity are also induced during the reactive stroma formation.

*Twist*, a helix–loop–helix transcription factor initially described in *Drosophila* (Ip *et al.*, 1992), was subsequently implicated in limb morphogenesis and EMT during embryonic development (Zuniga *et al.*, 2002). Twist proteins are essential for proper gastrulation, mesoderm formation, and neural crest migration during development, representing a typical EMT event. Within the context of cell survival, *Twist* functions as an anti-apoptotic factor and also acts as a regulator of the EMT by inducing the expression of mesenchymal markers such as fibronectin and N-cadherin (Yang *et al.*, 2004). *Twist* function is considerably up-regulated in prostate tumors (as well as other malignancies), and this *Twist* induction has been positively linked to poor prognosis, high Gleason grade, and decreased E-cadherin expression in prostate cancer (Kwok *et al.*, 2005). Significantly enough, down-regulation or loss of function of *Twist* in androgen-independent prostate cancer cell lines leads to decreased cancer cell migration and invasion capabilities. *Twist* signaling can modulate the apoptotic machinery by increasing the BCL-2/BAX ratio (Dupont *et al.*, 2001), that may provide a molecular basis for the ability of *Twist* to confer therapeutic resistance to taxol and vincristine in bladder, ovarian, prostate, and nasopharyngeal tumors. Loss of *Twist* expression renders cancer cells more sensitive to anoikis and TNF-α induced apoptosis (Kucharczak *et al.*, 2003; Sosic *et al.*, 2003).

Activation of NF-κB signaling, a major activator of immune and inflammatory functions (Karin *et al.*, 2002; Helbig *et al.*, 2003), is heavily associated with tumorigenesis mainly by conferring apoptosis resistance to tumor cells. Indeed NF-κB has been found to be constitutively activated in human breast, prostate, colorectal, and ovarian tumors (Huang *et al.*, 2001), hence the emergence of NF-κB as an ideal therapeutic target in cancer treatment (Orlowski and Baldwin, 2002). NF-κB in conjunction with Ras protects breast tumor cells from undergoing TGF-β-induced apoptosis, an important protective action that allows epithelial cells to proceed with the EMT process, essentially attributed to the ability of NF-κB to maintain these cells in the mesenchymal state (Ozes *et al.*, 1999). Mechanistically TGF-β signals in a Smad-independent fashion through TGF-β activated kinase 1 (TAK1), a kinase that phosphorylates IKK complex (Schiemann *et al.*, 2002). A considerable body of evidence implicates the NF-κB signaling as contributing to anoikis outcomes within the tumor microenvironment, with key anoikis regulators suppressing apoptosis (Shukla and Gupta, 2004), via activation of the PI3/AKT signaling cascade (Romashkova and Makarov, 1999). Since long-term androgen ablation has been shown to cause resistance to PI3K/AKT pathway inhibition in prostate tumors (Pfeil *et al.*, 2004), investigators have turned their efforts to determining the NF-κB status of prostate cancer cells prior to selecting the apoptotic route to be impaired (Rokhlin *et al.*, 2000). During the EMT process, epithelial cells acquire mesenchymal characteristics, and survival pathways such as the PI3/AKT signaling cascade are activated to ensure tumor mesenchymal cells successfully migrate in the stroma (Savagner, 2001). Tumor cell proliferation and invasion through the basement membrane into the stromal compartment causes a stromal response as an effort to repair the "damage". This response creates a new stromal microenvironment that is different from the native one because it shows ECM remodeling, elevated protease activity, increased growth factor bioavailability, increased angiogenesis, and influx of inflammatory cells (Sung and Chung, 2002). The "reactive stroma" enables an enhanced support system by providing a highly nurturing environment for the invading tumor cells (Rowley, 1998; Tuxhorn *et al.*, 2002a).

The solid hallmark characteristic of a reactive stroma is the presence of myofibroblasts, cellular intermediates between a fibroblast and a smooth muscle cell, which are characterized by a unique cytoskeletal protein expression profile and ultrastructural features (Gabbiani and Majno, 1972; Tlsty and Hein, 2001). A direct correlation between myofibroblasts and invasive metastatic carcinomas has been reported in several human malignancies including breast, lung, colon, stomach, and prostate cancer (Seemayer *et al.*, 1979). Myofibroblasts are directly involved in ECM remodeling since they secrete key ECM molecules such as fibronectin (embryonic fibronectins), collagen I and III, glycoproteins (tenascin and thrombospondin-1), and proteoglycans (Ibrahim *et al.*, 1993; Roberts, 1996; Ricciardelli *et al.*, 1998). During prostate tumorigenesis the progression to a highly aggressive malignant state is linked to the myofibroblast component of the prostate gland (Rowley, 1998; Tuxhorn, Ayala and Rowley, 2001; Tuxhorn *et al.*, 2002c). TGF-β is an active participant in the process by upregulating two growth factors, IGF-1 and HGF, both significant players in EMT (Tuxhorn *et al.*, 2002b).

In addition to ECM remodeling, the reactive stroma may also play a primary role in recruiting endothelial cells towards tumor neovascularization and metastasis to the bone (Sikes *et al.*, 2004). In the stromal component of breast carcinoma tissue specimens, a number of myofibroblasts can contribute to tumor cell proliferation, as well as to the recruitment of endothelial progenitor cells (EPC), creating highly vascularized tumors in mouse xenograft models (Franck-Lissbrant *et al.*, 1998; Myers *et al.*, 1999). These myofibroblasts (known as carcinoma-associated fibroblasts) are capable of attracting EPC through the release of high levels of SDF-1 that functions as a chemotractant to turn EPCs into carcinomas (Seemayer *et al.*, 1979; Orimo *et al.*, 2005). Overexpression of CXCR4 in various types of cancer xenograft models promoted tumor growth (Smith *et al.*, 2004), and targeting this signaling by either knockdown of CXCR4 expression (Sun *et al.*, 2003) or by small-molecule inhibitors, leads to tumor suppression. The SDF-1/CXCR4 axis has also been functionally involved in prostate tumor invasive behavior. Neutralization of the SDF-1 axis in prostate cancer cells halts their metastatic potential and also prevents their growth in the bone (Russell, Bennett and Stricken, 1998; Shah *et al.*, 2002). Stimulation by SDF-1 induces the expression of key cytokines such as IL-6 and IL-8, and growth factors such as VEGF, potentiating the vascularization of the tumors (Shariat *et al.*, 2001). Remodeling of the ECM by proteases is also important in the reactive stroma formation. Serine proteases and matrix metalloproteins (MMPs) have their expression increased, correlating positively with the tumor metastatic potential (Handsley and Edwards, 2005). The increased proteolytic activity seen in tumor tissues is a reflection of an imbalance between the proteases (MMPs) and their inhibitors (TIMPs — tissue inhibitor of metalloproteases). In prostate tumors of low Gleason grade, expression of TIMPs is higher compared to MMPs, while in higher Gleason tumors the reverse correlation is observed (Jung *et al.*, 1998). The reactive stroma might be responsible for the production of key molecules that not only assist in anoikis and apoptosis (Ramachandra *et al.*, 2002), but are also closely involved in prostate tumor neovascularization (Schor, Schor and Rushton, 1988; Shimura *et al.*, 2000), basement membrane organization, and tumor cell invasion (Rabinovitz *et al.*, 2001). In both modes of cell death, activation of survival pathways and EMT induction may characteristically occur in a concomitant fashion among the epithelial, endothelial, and stroma smooth muscle cell populations, rendering the dissection of the precise role of anoikis resistance as a critical event in prostate tumor progression a challenging task.

# Novel Molecular Therapeutics for Targeting Castration-Resistant Prostate Cancer

## 7.1 Therapeutic Targeting of TGF-β Signaling

TGF-β has been crowned as the "Director General" of the malignant process (Teicher, 2001; Akhurst and Derynck, 2001), a title justified by a wealth of evidence in the literature. Unquestionably, TGF-β stands out prominently among other cytokines due to its profound apoptotic, cytostatic, and migratory actions in normal homeostasis and cancer development (Siegel and Massague, 2003). In normal prostate epithelial cells and pre-malignant lesions (high-grade intraepithelial neoplasia), TGF-β plays a positive role as a tumor suppressor; thus targeting TGF-β too early in the tumorigenic process might generate the possibility of preventing its tumor-suppressing activity, so enabling rapid tumor growth (Akhurst and Derynck, 2001; Zhu and Kyprianou, 2005). However, one must also consider that elevation of TGF-β expression will quickly drive tumor progression, invasion, and metastasis; thus the temporal events responsible for the transition from growth suppressor to growth promoter must be characterized. The intricate details of the TGF-β transduction network that signal growth-inhibitory versus growth-promoting responses has been extensively investigated (Bachman and Park, 2005). Members of the TGF-β signaling family are being considered not only as predictive biomarkers of biochemical progression in prostate cancer patients (Kattan *et al.*, 2003; Shariat *et al.*, 2004a), but also as attractive molecular targets for the prevention and treatment of metastatic prostate cancer (Jakowlew, 2006).

In advanced prostate cancer, there is elevated expression of TGF-β with simultaneous loss of expression of TβRI and TβRII receptors. TGF-β overproduction enhances the metastatic ability of prostate tumor cells and provides a driving mechanism for the high tumor vascularity (Morton and Barrack, 1995). Both the TβRI and TβRII receptors are required to maintain the functional integrity of TGF-β signaling, and loss of either of these receptors results in resistance to TGF-β, permitting the cells to escape the growth constraints imposed by TGF-β. During tumor progression, TGF-β contributes to the metastatic process by promoting ECM degradation, invasion, and also affecting immunity (Thomas and Massague, 2005). Malignant cells become resistant to TGF-β growth-inhibitory effects potentially via mutations in cell cycle regulators such as p15, or alterations in the activation of Smad-independent pathways, such as pI3K and Ras, with the Smad-dependent signaling having a gradually diminishing role in the TGF-β growth regulatory network (Derynck and Zhang, 2003; Zhou *et al.*, 2010).

Elevated TGF-β levels in a patient's serum provides a significant prognostic value for highly aggressive metastatic disease and poor patient prognosis (Biswas *et al.*, 2006). Insightful investigation of a possible biomarker value such as TGF-β level, its signaling regulators, effectors, and receptors in prostate cancer progression may provide drug-targeting leads.

Current knowledge of the therapeutic targeting of TGF-β signaling will now be discussed. Pre-clinical studies on direct targeting of TGF-β (ligand) using anti-sense approaches and antibodies, and indirect inhibition of its membrane receptors, provide promise for the potential therapeutic value of targeting TGF-β signaling during tumor progression (Biswas *et al.*, 2006). TGF-β anti-sense treatments have been shown to reactivate tumor-specific immune responses in a variety of cancers (Iyer *et al.*, 2005). Anti-sense pharmaceutical companies recently focused their efforts on the development of a TGF-β-specific phosphorothioate anti-sense oligonucleotide, AP 11014, which is in the advanced stages of pre-clinical development as a potential treatment modality for advanced prostate cancer (Podar, Raje and Anderson, 2007). Schlingensiepen and colleagues reported that AP 11014 inhibited tumor cell proliferation, reduced migration, and reversed immunosuppression in lung, colorectal, and prostate cancers (Schlingensiepen *et al.*, 2004*)*. Anti-sense pharmaceutical companies have enjoyed a relatively high degree of success in the anti-TGF-β2 oligonucleotide AP 12009 in the treatment of glioma, pancreatic, and malignant melanoma cell lines, which are in phase I/II studies (Podar, Raje and Anderson, 2007). AP 12009 has proven to be well-tolerated by patients, with few negative side-effects, providing promise for its therapeutic value.

TGF-β antibodies function in a similar pattern to the TGF-β anti-sense oligonucleotides, that is, towards bypassing the immunosuppressive effects of TGF-β. They have achieved a relative degree of success in trial-based therapeutic attempts in several cancers. Direct targeting of TGF-β receptors also provides a promising therapeutic avenue for clinical exploitation. Blocking the signaling cascade at its origin would impair activation of the downstream intracellular effectors, consequently abrogating the anti-growth actions of TGF-β (Nijtmans *et al.*, 2000). Elegant *in vivo* studies have demonstrated that disruption of TβRII signaling using a dominant negative receptor (DNRII) leads to suppression of metastasis in a mouse model (Iyer *et al.*, 2005). Introduction of a DNRII to bone marrow cells leads to leukocyte formation capable of anti-tumor activity and suppression of metastasis in prostate cancer models (Kaminska, Wesolowska and Danilkiewicz, 2005). Functional inhibition of TGF-β activities can be accomplished by the use of agents such as stabilized soluble protein TβRII:Fc, that significantly reduce metastases without any major side effects (Yang *et al.*, 2004). Systemic administration of sBG has been shown to inhibit prostate and breast cancer growth and angiogenesis in xenograft models (Bandyopadhyay *et al.*, 2004). Additional data regarding the direct therapeutic targeting of TGF-β signaling stem from gene therapy-based experimental trials in mice, documenting substantial suppression of tumor metastasis by impairing TGF-β signaling in bone marrow cells (Shah *et al.*, 2002).

Studies on the development of small molecule inhibitors acting as TβRI kinase inhibitors indicate considerable promise in the ability of these agents to inhibit the proliferative and immunosuppressive effects of TGF-β in cell-based assays and in cancer models (Biswas *et al.*, 2006). Thus LY215799, a dihydropyrolopyrazole derivative, is a selective TGF-βRI kinase inhibitor that inhibits Smad signaling in the TGF-β intracellular pathway (Pinkas and Teicher, 2006). A comprehensive report by Sawyer concentrates on the efforts by Eli Lilly Research to test many other small molecule inhibitors, such as LY550410, LY580276, and LY364947 (Yingling, Blanchard and Sawyer, 2004). These compounds are competitive inhibitors of the ATP binding site of TβRI kinase. Issues related to the stability, rapid clearance, specificity, and side effects of these receptor inhibitors, however, raise reasonable concerns. A significant side effect that merits attention as the drug-optimization process expands is associated with the targeting/inhibition of the tumor suppressor function of the TGF-β receptors, loss of which has been correlated with biochemical progression of patients undergoing radical prostatectomy (Shariat *et al.*, 2004a). Nevertheless, the knowledge gathered from molecular insights into novel effectors so far would allow for the combination targeting of TGF-β, towards suppression of not only the primary tumor growth, but also metastatic spread (Akhurst, 2002; Jones, Pu and Kyprianou, 2008). The apoptotic action of doxazosin and terazosin (quinazoline-based $\alpha_1$-adrenoreceptor antagonists), which triggers the phenomenon of anoikis in prostate cancer cells, could be consequential to targeting TGF-β receptors (Benning and Kyprianou, 2002), a mechanism with a potential therapeutic promise for impairing metastasis initiation.

The characterization of the signaling network operated by TGF-β, via Smad-dependent and Smad-independent mechanisms in conjunction with branches of MAP kinase pathways, has enabled the therapeutic exploitation of this multifaceted growth factor during cancer development and progression. Through synergistic collaboration with the NF-κB, JNK and Ras signaling pathways, TGF-β can activate IL-6 expression in prostate tumors (Park *et al.*, 2003). The challenge remains to identify effective therapeutic modalities that do not interfere with the positive effects of TGF-β as a tumor suppressor at the onset of malignant transformation and tumorigenic growth, while inhibiting the dysfunctional TGF-β in advanced prostate cancer. Considering this ill-fated

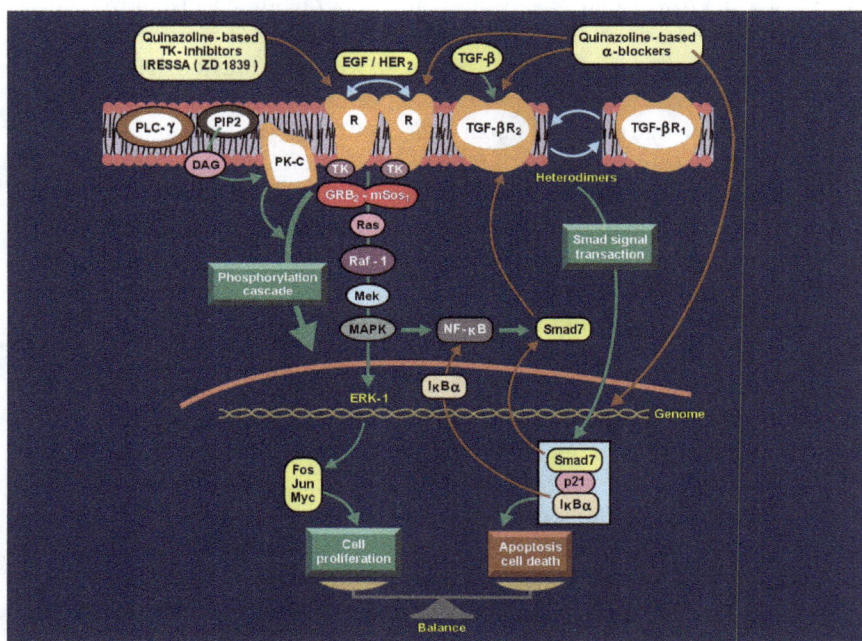

**Figure 7.1**   A model outlining the potential action of quinazoline-based α₁-adrenoceptor antagonists on apoptosis signaling pathways in prostate cancer cells. TGF-β signaling involves formation of the receptor complex, Smad activation, nuclear translocation, and gene transcription subsequently resulting in apoptosis. Cell survival or mitogenic pathways initiated by activated protein tyrosine kinases (PTK) (in response to EGF) can repress this pathway at different levels and may inhibit TGF-β signaling. Quinazoline-based antagonists induce apoptosis via internalization with TGF-β receptors and by engaging the TGF-β signaling pathway and Iκ-B activation towards apoptosis induction (Partin, Anglin and Kyprianou, 2003). This is a comparative mechanism similar to the action of quinazoline-based TK-inhibitors (IRESSA) targeting the EGF/HER2 pathway.

alliance between TGF-β and tumors, the temporal targeting of the TGF-β pathway should be implemented during the transition from tumor suppressor to tumor promoter, and prior to its involvement in the metastatic process in the context of the tumor microenvironment. The biological repertoire of the epithelial and endothelial cells under the control of TGF-β signaling is intimately associated with the anoikis phenomenon towards prostate tumor metastasis. Based on the pre-clinical trials, blocking TGF-β signaling provides a potentially potent and novel approach to treating patients with metastatic prostate cancer. Investment in neutralizing antibodies as selective therapeutics targeting signal transduction pathways has also provided support for treatment outcomes in the clinic. Moreover, there is intense clinical interest in the efficacy of systematically administered short hairpin RNA (shRNA) as an effective therapeutic strategy in several human cancers (Grimm *et al.*, 2005). Considering the regulatory role of p53 in mediating TGF-β response, this may provide an alternative therapeutic platform via indirect targeting of TGF-β signaling (Piccolo, 2008). Identification of a predictive marker on a molecular signature before or during pre-clinical anti-tumor drug development will enable the selection of patients who are likely to respond and exhibit therapeutic benefit versus non-responders.

The functional interplay with other growth factor signaling pathways, such as fibroblast growth factor (FGF) (Nakano *et al.*, 1999; Rosini *et al.*, 2002), insulin-like growth factor-1 (IGF-1) (Miyake *et al.*, 2000; Song *et al.*, 2003), epidermal growth factor (EGF) (Orio *et al.*, 2002), and vascular endothelial growth factor (VEGF) (Haggstorm *et al.*, 1998; Breier *et al.*, 2002), may impact the regulatory involvement of TGF-β in angiogenesis and metastasis, and consequently the outcome of therapeutic targeting. Combination approaches targeting multiple growth factor pathways and frequently overlapping signaling effectors may provide the ultimately effective modality for treating patients with advanced metastatic prostate cancer in a "personalized medicine" approach. Dissection of the interaction of TGF-β and AR can lead to new therapeutic options based on inhibition of vital interactions/cross-talk between TGF-β, its receptors and AR (Chipuk *et al.*, 2002; Wang *et al.*, 2005; Zhu and Kyprianou, 2008; Knudsen and Scher, 2009). Ongoing studies suggest a combination of targeting of TGF-β signaling with novel therapeutic compounds as AR antagonists.

## 7.2 Exploitation of Quinazolines: Lifting Anoikis Resistance to Impair Metastasis

The primary endogenous inhibitor of the death receptor pathway is the FLICE-inhibitory protein (FLIP). This effector consists of two DEDs and a C terminus caspase-like domain which provide the structural similarity to caspase-8. Upon binding, FLIP substitutes active-site cysteine bond to tyrosine, leading to inactivation of caspase-8. Functionally FLIP has higher affinity to the DISC compared with caspase-8, thus blocking caspase-8 recruitments and subsequent activation (Scaffidi *et al.*, 1999). In normal cells, FAS ligands and FAS receptor are up-regulated, while FLIP expression is down-regulated upon loss of cell contact with ECM. These changes in apoptosis signaling effectors trigger downstream caspase-8 activation in a FADD-dependent manner, leading to apoptosis (Aoudjit and Vuori, 2001). In the prostate tumor micro-environment, even though epithelial and endothelial cells lose contact with the ECM, the system becomes dysfunctional as tumor epithelial cells fail to activate the death receptor-mediated caspase-8 activation, despite increased expression of FAS ligands and FAS receptors, ultimately conferring anoikis resistance (Sakamoto and Kyprianou, 2010). The driving molecular mechanism of this defect can be partially explained by the failure of cancer cells to down-regulate FLIP expression after detachment (Simpson, Anyiwe and Schimmer, 2008). FLIP's functional requirement to maintain prostate tumor apoptosis resistance provides the first basis for targeting FLIP in anoikis-resistant cells towards the formulation of a novel therapeutic approach to treat metastatic prostate cancer (Zhang *et al.*, 2004a). The emerging evidence is very promising: using small molecules, chemical inhibition of FLIP sensitized cells to apoptosis stimuli, and reversed the anoikis resistance in malignant cells, without inducing apoptosis in adherent cells (Mawji *et al.*, 2007a; Mawji *et al.*, 2007b).

Experimental and clinical evidence gathered by our group and others indicates that two clinically available quinazoline-based $\alpha_1$-adrenoceptor antagonists, doxazosin and terazosin, exert potent anti-growth effects via induction of prostate epithelial cell apoptosis, in addition to causing smooth muscle relaxation and a decrease in vascular pressure (Chon *et al.*, 1999; Benning and Kyprianou, 2002; Walden, Globina and Nieder, 2004). The observation that clinically available FDA-approved drugs selectively induce apoptosis via an anoikis effect, without affecting cell proliferation of prostate epithelial and endothelial cells, has profound translational ramifications in the medical management towards a cure of prostate cancer patients with metastatic disease (Tahmatzopulos and Kyprianou, 2004; Harris *et al.*, 2007). Suppression of prostate tumor cell growth by these drugs proceeds via an $\alpha_1$-adrenoceptor-independent mechanism, mediated by induction of the TGF-$\beta$ apoptotic signaling upon plasma membrane internalization facilitated by TGF-$\beta$ receptors T$\beta$RI and T$\beta$RII, and by downstream inhibition of NF-$\kappa$B activation (Partin, Anglin and Kyprianou, 2003) (Fig. 7.1). Subsequent studies revealed that the final execution of the cells proceeds via death receptor-mediated apoptosis involving DISC formation/caspase-8 activation, and inhibition of AKT activation (Garrison and Kyprianou, 2006). After this initial demonstration of the apoptotic action of doxazosin and other quinaline-based $\alpha_1$-adrenoceptor antagonists, our efforts focused on the structural optimization of doxazosin's quinazoline nucleus, and led to the development of novel compounds with potent apoptotic and anti-vascular activity by targeting AKT-survival signaling (Shaw *et al.*, 2004; Garrison *et al.*, 2007). The events involved in tumor progression to metastasis are largely mediated by the integrins, which upon engagement with components of the ECM, reorganize to form adhesion complexes (Fornaro, Manes and Languino, 2001). As discussed earlier, anoikis represents an exotic mode of cell death, and apoptosis consequential to insufficient cell-matrix interactions plays a prominent role in prostate tumor angiogenesis and metastasis. The goal is to maximize the therapeutic value of new quinazoline-derived agents to overcome anoikis resistance (by AKT activation), and suppress prostate tumor invasion and metastasis for treatment of advanced prostate cancer.

The separation of doxazosin's effect on cancer cell apoptosis from its original pharmacological activity in smooth muscle cells provided an ideal platform for lead optimization of apoptosis-inducing agents. Recent structural optimization of quinazoline's chemical nucleus and structure–function analysis led to the development of a novel class of apoptosis-inducing and angiogenesis-targeting agents (Shaw *et al.*, 2004). Two lead compounds, DZ-3 and DZ-50, are effective at reducing endothelial cell viability and preventing *in vitro* tube formation and *in vivo* vessel development respectively. DZ-50 also significantly reduced tumor cell ability to attach to ECM and migrate through endothelial monolayer via anoikis. *In vivo*, this drug leads to significant suppression of tumorigenic growth, as well as prevention of tumor initiation, in androgen-independent human prostate cancer xenografts. Moreover, DZ-50 treatment reduced formation of prostate tumor-derived metastatic lesions. Two new quinazoline-based compounds, DZ-3 and DZ-50, have higher potency than doxazosin in suppressing prostate growth by targeting apoptosis (DZ-3) and vascularity (DZ-50), at lower doses, thus providing therapeutic promise for metastatic prostate cancer. Integrin $\beta_1$ deregulation in response to DZ-50 might be a consequence of alterations in the focal adhesion complex (talin, FAK), and key components of the actin microfilaments that determine cell motility. Dissecting the ability of quinazolines to target the interactions between integrin $\beta_1$ with its intracellular signaling partners, such as talin, is highly significant, as by reducing the migratory capacity of tumor cells and/or endothelial cells via anoikis, we could effectively prevent metastasis and angiogenesis (Garrison and Kyprianou, 2006; Garrison *et al.*, 2007).

## 7.3 Receptor Tyrosine Kinase Targeting

Among the known receptor tyrosine kinases (RTKs), the epidermal growth factor receptor (EGFR) and Erb-2 (HER-2) are the two prototypic family members. Overexpression of EGFR and Erb-2 has been documented as a predictor of aggressive disease and poor patient prognosis in a range of human malignancies including breast and prostate cancer (Craft *et al.*, 1999; Hernes *et al.*, 2004). Elevations in EGFR expression occur in prostate cancer cells and associated endothelial cells from bony metastases, as opposed to other metastatic sites in experimental xenograft models (Kim *et al.*, 2003). In human prostate tumor specimens, elevated EGFR expression correlates with increased stage, Gleason grade, PSA, invasion, and progression to metastasis (Di Lorenzo *et al.*, 2002; Hernes *et al.*, 2004; Shuch *et al.*, 2004). Most significantly it has been associated with the molecular switch to androgen independence (Hernes *et al.*, 2004) and has been implicated in the racial disparities existing in prostate cancer disease behavior and outcomes (Shuch *et al.*, 2004). This constellation of data provided a strong molecular basis for the development of inhibitors of the EGF-EGFR pathway that can induce cell cycle-independent death of prostate tumors (Dionne *et al.*, 1998) and target androgen-independent prostate cancer with metastatic lesions to the bone (Kim *et al.*, 2003), by impairing tumor-associated endothelial cell growth. Various approaches have been developed during the last decade to target the EGFR signaling pathway including mono-clonal antibodies directed against the receptor and synthetic tyrosine kinase inhibitors (IRESSA and TARCEVA). As shown in Fig. 7.1, disruption of these kinases has an anti-proliferative effect against tumor cells. Although this has been the classic understanding of the action of these drugs, studies with experimental models established that disrupting the EGF-EGFR signaling can also lead to apoptosis induction in prostate cancer cells (Harper *et al.*, 2002; Kim *et al.*, 2003; Farhana *et al.*, 2004). The pre-clinical data led to clinical trials with compounds such as the EGFR tyrosine kinase inhibitor IRESSA. The early results in prostate cancer patients have met with a degree of variability and thus therapeutic uncertainty; nevertheless it must be recognized that disruption of this critical signaling mechanism represents a therapeutic target of profound dimensions and significance (Blackledge, 2003). Despite years of intensive research on EGFR inhibitors, there is a surprising dearth of chemically distinct small inhibitors with a high degree of selectivity. A need also arises for new inhibitors due to the recent findings of EGFR mutations which render the kinases resistant to gefitinib.

## 7.4  Histone Deacetylase Inhibitors (HDACs): Therapeutic Inhibitors

Inhibitors of histone deacetylases (HDACs) have emerged as potent therapeutic agents for the treatment of solid tumors and hematological malignancies (Marks and Dokmanovic, 2005). These agents trigger growth arrest, differentiation, and apoptosis in tumor cells via transcriptional activation of certain genes and potential inactivation of tumor-suppressor genes (Bolden, Peart and Johnstone, 2006). The acetylation status of core histones plays an important role in regulating gene transcription through the modulation of nucleosomal packaging of DNA. Histone acetylation leads to relaxed nucleosomal structures, giving rise to a transcriptionally permissive chromatin state. The level of this post-transcriptional modification is maintained by a dynamic balance between the activities of histone acetyltransferases (HATs) and histone HDACs, both of which are recruited to target genes in complexes with sequence-specific transcriptional activators. There is a characteristically high expression of HDAC isoenzymes and a corresponding hypoacetylation of histones found in cancer cells (Minucci and Pelicci, 2006). Moreover, loss of genetic gatekeeper function in pre-cancerous lesions may be associated with increased activity of HDACs. HDAC inhibitors have shown pleiotropic anti-tumor activities in recent clinical and pre-clinical investigations of human cancers, both through histone acetylation-dependent and through histone acetylation-independent mechanisms. Given the vast biologic effects of HDAC inhibition, one might expect HDAC inhibitors to have a narrow therapeutic window; however, one must recognize that HDAC inhibitors induce the activation of genes silenced in tumor growth through chromatin remodeling (Marks, Richon and Rifkind, 1996).

In the case of prostate cancer, aberrant epigenetic regulation by HDAC plays a dominant role in the etiology and progression to metastasis. There is compelling evidence to suggest that elevation of the HDAC level drives the transformation of normal prostate epithelial cells via activation of various oncogenes and repression of key tumor suppressor genes, ultimately blocking apoptosis induction (Halkidou *et al.*, 2004; Nakagawa *et al.*, 2007; Weichert *et al.*, 2008). Thus HDAC down-regulation offered wide promise as a critical therapeutic target leading to reversion of malignant transformation and is thus potentially effective in prostate cancer treatment. Several structurally distinct HDAC inhibitors have entered phase I or II clinical trials for human cancer, including prostate cancer, and the promise invested in their therapeutic value is reflected by the recent FDA approval of vorinostat (Marks and Dokmanovic, 2005). Recent studies on the structural optimization of short-chain fatty acids by coupling the Zn-chelating motifs through aromatic linkers were conducted at Ohio State University by Dr Chen and his team. These led to the development of novel phenylbutyrate-based HDAC inhibitors that exhibit potent apoptosis-driven anti-cancer effects, and possible preventive action impairing prostate cancer development and progression to metastasis (Sargeant *et al.*, 2008). This new class of HDAC inhibitors suppresses prostate tumor growth and vascularity by targeting the AKT, surviving, and BCL-XL survival signaling pathways (Kulp *et al.*, 2006). The future for these molecules holds additional promise due to the differential sensitivity of the transformed cells to their action compared to the normal cells, and consequently to their minimal toxicity *in vivo*.

In an interesting mechanistic twist, it was recently reported that HDAC inhibitors, including SAHA (vorinostat) and LBH589, which are currently being tested in clinic, could provide alternative therapeutic options by targeting

AR-transcriptional activity in CRPC (Welsbie *et al.*, 2009). Specifically, HDAC inhibitors were shown to block the AR-mediated transcriptional activation of the TMPRSS2 gene involved in fusion with ETS family members in a majority of prostate cancers. In response to HDAC inhibitor treatment there was a marked reduction in AR protein levels in prostate cancer cells; however independently of AR protein content, HDAC inhibitors block AR activity through inhibiting the assembly of co-activator/RNA polymerase II complex after AR binding to target gene enhancers. The ability of HDAC inhibitors to block AR activity in CRPC models merits clinical investigation towards an effective therapeutic strategy for CRPC, with the HDAC-regulated AR target genes serving as potential biomarkers.

## 7.5 Selective Death Action by Cancer-Specific PAR-4 in Prostate Tumors

Prostate apoptosis response-4 protein (Par-4) is a leucine zipper domain protein that is conserved in vertebrates (Sells *et al.*, 1994). Like any "classic" tumor suppressor gene, Par-4 overexpression is sufficient to induce apoptosis in cancer cells, but unlike other tumor suppressors does not cause apoptosis in normal cells. The anti-tumor action of Par-4 proceeds via two distinct pathways: activation of molecular components of the cell death machinery and inhibition of pro-survival factors. The human Par-4 gene is located on the minus strand of chromosome 12q21.2, and encodes for a protein that is 38kDa in size, containing 342 amino acids. Par-4 was identified as a pro-apoptotic protein in the prostate in pioneering studies by Dr Vivek Rangnekar and colleagues and has been subsequently characterized as a tumor suppressor (Zhao and Rangnekar, 2008). In a striking fashion, Par-4 overexpression is sufficient to induce apoptosis in most cancer cells in the absence of a second apoptotic signal, but does not cause apoptosis in normal or immortalized cells. Par-4 has been found to be an essential downstream regulator of cell death programs initiated by various exogenous signals such as TRAIL, vincristine, and radiation (Sharifi *et al.*, 2007) Par-4 induces apoptosis by concomitantly activating FAS death receptor signaling and inhibiting NF-κB cell survival activity (Chakraborty *et al.*, 2001), while targeted ablation of Par-4 reveals a cell-type specific susceptibility to apoptosis-inducing agents (Affar *et al.*, 2006). The core effector domain of Par-4 (amino acids 137-195, designated the SAC domain) is necessary and sufficient to induce apoptosis (Gurumurthy *et al.*, 2005). The tantalizing feature of this molecule is the ability to induce apoptosis selectively in cancer cells, but not in normal cells, because endogenous Par-4 is inactive and its apoptotic potential is unleashed in response to apoptotic insults.

Mechanistically the story unfolds in a most intriguing manner: the cancer cell-specific apoptotic action of Par-4 and the SAC domain is attributed to their selective activation via phosphorylation at the T155 residue by protein kinase A (PKA) in cancer cells (Goswami *et al.*, 2005). Importantly, endogenous Par-4 is inactivated owing to phosphorylation by AKT1 in prostate cancer cells (Goswami *et al.*, 2005). The binding and phosphorylation of Par-4 by AKT makes Par-4 a substrate for the chaperone protein 14-3-3, which effectively sequesters Par-4 in the cytoplasm (Gurova *et al.*, 2005). Assessment of tumor formation in B6C3F1 mice indicates the development of spontaneous tumors in the spleen and liver at a high frequency in the control animals, whereas none of the Par-4- or SAC-expressing transgenic animals shows spontaneous tumors. In contrast, offspring of the Par-4- or SAC-expressing transgenic animals and TRAMP animals demonstrate mainly prostatic intraepithelial neoplasia (PIN) lesions. Importantly, the PIN lesions that progressed to adenocarcinoma had lost the Par-4 or SAC transgene. This implies that Par-4 and the SAC domain suppress the growth of prostatic tumors, and their loss is essential for prostate tumor growth. Although ectopic Par-4 and the SAC domain exert their apoptotic effect by acting within the cell nucleus to inhibit NF-κB, initial data identified a here-to-fore unrecognized secreted SAC domain (sSAC) that retains apoptotic activity (Zhao *et al.*, 2007). Thus, it could be that sSAC exerts cancer-specific apoptotic effects via extracellular mechanisms.

## 7.6 Death Synergy Between Proteosome and Death Receptor Leads to Tumor Regression

The 26S proteosome accounts for the bulk of cellular protein turnover (Glickman and Ciechanover, 2002). The proteolytic core of the 26S proteosome — the 20S proteosome — contains several peptidase activities including chymotrypsin-like, trypsin-like, and post-glutamyl activity. It is the 20S chymotrypsin-like activity that the clinically relevant proteosome-inhibiting compound targets (Kisselev and Goldberg, 2001). Velcade (bortezomib) is a highly bioavailable synthetic peptide boronate specifically designed to inhibit the proteosome. In pre-clinical studies, Velcade induces growth arrest and apoptosis in many tumor types *in vitro* (Adams *et al.*, 1999). Subsequent animal studies using tumor xenografts further demonstrated that Velcade could reduce tumor volume, and exhibited minor toxicity (LeBlanc *et al.*, 2002). Clinical trials conducted in multiple myeloma patients have validated the safety and therapeutic value of Velcade (Richardson *et al.*, 2005). Studies on the mechanism by which Velcade leads to cell death have focused on cell-intrinsic pathways, such as NF-κB signaling in cell culture models (Nalepa *et al.*, 2006). However, these studies explain only a fraction of the anti-cancer activity *in vitro* and fail to take into account the presence of other cells and factors in the tumor microenvironment that render cells susceptible to apoptosis through a cell-extrinsic program (Burger and Seth, 2004).

The clinical presence of the proteosome inhibitor bortzemib (PS-341) (Velcade) is rigorously established. The drug has indeed been approved as a first-line therapy for treating multiple myeloma, an aggressive hematopoietic malignancy caused by the clonal proliferation of plasma cells. Apoptosis induction has been implicated as the mechanism driving the therapeutic effect of proteosome inhibitors in these target cells, which can be blocked by caspase inhibitors. Pharmacologic inhibition of the 26S proteosome by Velcade (PS-341/bortezomib) is a viable chemotherapeutic strategy that effectively treats patients with refractory multiple myeloma (Adams, 2004). Clinical trials are currently under way to determine the efficacy of Velcade alone or in combination with other agents in multiple myeloma and hormone-refractory prostate cancer patients (Adams *et al.*, 1999). As determined in a phase III clinical trial, 6% of refractory multiple myeloma patients exhibited a complete response and an additional 39% a partial response to Velcade (Richardson *et al.*, 2005). While the prospect that this agent can also be an effective treatment for metastatic prostate cancer is encouraging, considering that 22–36% of patients with CRPC have shown a significant decrease in markers of disease progression (Papandreou *et al.*, 2004; Price and Dreicer, 2004), there is room for considerable improvement. Proteosome inhibitors have clearly proven their utility in the clinic; however, their development remains at an early stage. Improving the response rate in prostate cancer patients with advanced CRPC is a primary challenge in advancing proteosome-directed anti-cancer therapies as individual strategies or combination approaches (Papandreou and Logotheis, 2004; Thorpe *et al.*, 2008). The main obstacle to improving the efficacy of Velcade is that the mechanism(s) by which proteosome inhibition leads to tumor cell death are poorly understood. The majority of studies have focused on angiogenesis, ER stress, or cell-intrinsic changes that lead to apoptosis (Adams, 2004; Nalepa, Rolfe and Harper, 2006). In contrast, recent evidence established that proteosome inhibition completely sensitizes multiple prostate cancer cell lines to apoptosis mediated by the cell death cytokines FAS ligand and TNF-related apoptosis-inducing ligand (TRAIL) (Thorpe *et al.*, 2008), while normal prostate epithelial cells are resistant to this regimen. Moreover, *in vivo* studies demonstrated that the TRAIL/Velcade combination is well tolerated and results in a robust apoptosis induction, directly translating into an equally impressive anti-tumor response. Velcade and death receptor agonists (TRAIL, FAS ligand, or antibodies that activated the TRAIL receptors) have both met severe limitations in the treatment of cancer. Although Velcade has a high partial response rate (50%) the complete response rate is quite low (6%). Likewise, a majority of epithelial cancer cell lines, especially prostate, are completely or almost completely resistant to death receptor activation. Indeed, very low levels of Velcade are sufficient to remove the apoptosis "block" initiated by FAS ligand

and TRAIL in prostate cancer cells, both in culture and prostate tumor xenografts.

In direct accord with a functional interplay between proteosome inhibition and extrinsic apoptotic pathway control/activation, an additional mechanism for improving the therapeutic efficacy of Velcade has recently emerged: stabilization of the executioner caspases. Using a strikingly designed *in vitro* model, Gray and colleagues have recently reported a dose-dependent activation of specific caspases with a molecularly engineered small molecule-activated protease, an approach that might meet therapeutic needs in the clinic by reducing the apoptotic threshold in cancer cells (Gray *et al.*, 2010). Adopting this strategy in the clinic could improve the response rate of proteosome inhibitors by impairing a major apoptotic signaling pathway in synergy with proapoptotic agents activating caspases.

## 7.7 The SERCA Pump as a Therapeutic Target

The endoplasmic reticulum (ER) has emerged as an organelle that plays a major role in cell signaling pathways, cellular response to stress, and cellular activation of apoptosis (Ferri and Kroemer, 2001; Schwarze *et al.*, 2008). Consequently, therapeutic exploitation of ESR apoptotic pathways has recently been attempted for the treatment of advanced prostate cancer. The sarcoplasmic/endoplasmic reticulum calcium ATPase (SERCA) pump is identified as an ER protein whose normal function, to maintain the calcium gradient between the ER (mM $Ca^{2+}$) and the cytosol (nM $Ca^{2+}$), is required by all cells regardless of their proliferative state (Furuya *et al.*, 1996). The SERCA pump is present in all cell types, serving the key role of transferring $Ca^{2+}$ from the cytosol to the sarcoplasmic lumen, and inhibition of this pump action often leads to ER $Ca^{2+}$ depletion and elevation of cytoplasmic $Ca^{2+}$ (Shen *et al.*, 2002). Thapsigargin, a known pharmacological inducer of ER stress and an effective apoptosis-inducing agent, inhibits the SERCA pump. Indeed, a number of putative pharmacologic agents have been described that can induce ER stress, including blocking ER to Golgi transport (brefeldin A) and disrupting ER $Ca^{2+}$ stores through inhibition of SERCA pumps (thapsigargin). Depletion of ER $Ca^{2+}$ levels activates ER-stress response and, if sustained, triggers apoptosis via initial activation of intrinsic pathway involving ER-resident caspases, and subsequent elevation of cytoplasmic $Ca^{2+}$, which can activate the extrinsic apoptotic pathway (Schwarze *et al.*, 2008). In order to control targeting specificity and prevent bystander cytotoxicity, a PSA-cleavable, inactive thapsigargin derivative was synthesized (Denmeade and Isaacs, 2005). In theory, proteolysis in PSA-producing sites will hydrolyze this pro-drug into an active bioavailable agent at prostate tumor sites. Thus the ER functional integrity and in particular the SERCA pump represents an attractive apoptosis-driven therapeutic target for cancer therapy. While the role played by calcium release in this process remains controversial, what has been recognized is that mechanisms directing inhibition of the SERCA pump lead to apoptosis induction, becoming a favorable route for targeting the slowly-proliferating prostate cancer cells (Denmeade and Isaacs, 2005). Thus sustained SERCA inhibition by agents such as thapsigargin results in activation of ER-stress response and simultaneous activation of apoptotic pathways within the ER and the mitochondria (Denmeade and Isaacs, 2005).

There is a striking drawback, however. Given the SERCA pump's critical role in normal cellular metabolism, agents such as thapsigargin directed towards inhibiting SERCA function would be likely to produce significant toxicity to normal epithelial cells. The National Cancer Institute's anti-cancer screens revealed that thapsigargin has a broad growth-suppressing activity with an effective potency 10–100 times higher than the paclitaxel used in treating prostate cancer patients. Therefore, thapsigargin's selective targeting to prostate tumor sites became a requirement before its further pharmacologic exploitation as a therapeutic agent could be progressed. To that effect, ground-breaking studies by Isaacs and his colleagues established the anti-tumor action of thapsigargin can be attenuated by coupling to a targeting peptide to produce an inactive prodrug that is only activated by prostate cancer-specific proteases such as the serine protease prostate-specific antigen (PSA) (Jacobsen, Girman and Lieber, 2001; Denmeade and Isaacs, 2005). PSA-activated thapsigargin prodrugs have been characterized that are selectively toxic to PSA-producing prostate cancer cells *in vitro* and *in vivo* (Denmeade and Isaacs, 2005). The PSA-activated thapsigargin prodrug approach also overcomes the problem of heterogeneity in the production of target protease PSA by individual prostate epithelial cells in a given metastatic site. Considering that the extracellular fluid of human prostate cancer metastases contains high levels of enzymatically active PSA, a substantial apoptotic bystander effect is expected, which can be therapeutically beneficial for the metastatic lesions (Denmeade and Isaacs, 2005). These prodrugs are under intense pre-clinical evaluation as potential targeted therapy for prostate cancer. Although the use of PSA to cleave prodrugs may allow the induction of ER stress-induced apoptosis in prostate cells, no specific biochemical activity, biological process, or prostate-specific gene has yet been identified to suggest that targeting prostate cancer cells through ESR pathways may have high therapeutic value compared to other apoptotic routes. The limitation is that the UPR has been characterized best in yeast and in the mouse model. Due to the evolution of the ER stress pathway, the exact biochemical pathways that each ER stress gene activates in human cells must be characterized. In addition, while prostate tumors are highly heterogeneous, they are primarily composed of epithelial cells and one has to consider that much of the mammalian ESR research has exploited fibroblast models that may not accurately reflect pathways utilized in prostate cancer cells. For example, studies on mouse plasma cell differentiation demonstrated a function of the transcription factor XBP-1 as a molecular intersection of unfolded protein response, unknown in fibroblasts (Iwakoshi *et al.*, 2003). Considering that mice lacking ATFα and β, IRE1α, and XBP-1 are not viable, one must assume that these players have multiple vital functions in other cell types that are yet to be identified. The challenge will be to dissect the role of the individual stress sensors in the process of prostate tumor progression. This could allow the design of drugs which will selectively target the proper ESR regulatory molecules, with the therapeutic promise recently shown by the targeting of the extrinsic death receptor pathway (Garrison and Kyprianou, 2006; Garrison *et al.*, 2007).

## 7.8 Endothelin-Receptor Antagonists

Endothelins (endothelin-1 (ET-1), endothelin-2 (ET-2), and endothelin-3 (ET-3)) are regulators of cell proliferation, vasomotor tone, and angiogenesis. The endothelins bind to two receptors, endothelin-A and endothelin-B, and play a major role in the regulation of tumor growth, cell proliferation, apoptosis, angiogenesis, and bone metastasis during prostate cancer development and progression (Thakkar, Choueiri and Garcia, 2006). Pioneering work by Joel Nelson and his group provided robust evidence to document that patients with metastatic prostate cancer have elevated plasma endothelin-1 compared with patients with organ-confined disease. At the cellular level endothelin-A promotes osteoblastic activity characteristic of bone metastases in prostate cancer (Nelson *et al.*, 1995). These seminal findings provided a robust rationale for the clinical development of endothelin receptor antagonists to target metastatic prostate cancer. Atrasentan is predominantly an endothelin-A receptor antagonist that was extensively studied in two phase III trials (Carducci *et al.*, 2007). The initial clinical results have been rather disappointing, as atrasentan did not reduce the risk of disease progression relative to the placebo (according to radiographic and clinical measures), despite the fact that PSA levels were significantly lower in the atrasentan arm. The tolerance profile for atrasentan is relatively good, and the most common adverse events associated with treatment are headache, rhinitis, and peripheral edema, reflecting the vasodilatory and fluid-retention properties of the endothelin-A receptor blockade. Ongoing phase III clinical trials focus on the evaluation of atrasentan in combination with docetaxel/prednisone as a first-line treatment in metastatic CRPC.

Zibotentan (ZD4054) is a selective endothelin-A receptor antagonist that was reported in phase III trials to have similar results to atrasentan. However, an interim analysis revealed an improvement in OS (23.5, 24.5 versus 17.3 months for placebo). Built on these encouraging results, three phase III trials (ENTHUSE) involving a large number of patients with CRPC (2500) are currently being conducted with ZD4054 (Vishnu and Tan, 2010).

## 7.9 The Power of Sex Steroid Targeting

Cholesterol, originating from acetic acid towards steroid hormone synthesis, is the basic substrate fundamental for the synthetic steroidogenesis pathway. The most powerful intracellular intraprostatic androgen, 5α-dihydrotestosterone (5α-DHT), is formed through the standard steroidogenesis pathway, primarily from the plasma-derived testosterone (Bruchovsky and Wilson, 1968). Alternatively, benign and malignant prostate cells have the ability to convert adrenal-derived steroids such as androstenediol and dihydro-epinandrosterone (DHEA) to testosterone, due to increased expression of genes encoding for the enzymes converting adrenal androgens to testosterone (17-β hydroxy dehydrogenase and 3β-hydroxysteroid dehydrogenase, respectively) (Stanbrough et al., 2006). Sex steroid hormone signaling has been aggressively targeted by a number of drugs, including those approved by the FDA, developed for the purpose of treating benign and malignant prostate diseases (Marks, Mostaghel and Nelson, 2008), or for the treatment of breast cancer, osteoporosis, and other conditions. Significantly enough, different investigative groups have documented that substantial amounts of testosterone and DHT are present in human prostatic tissue after androgen deprivation therapy (Mohler et al., 2004; Titus et al., 2005; Page et al., 2006; Montgomery et al., 2008). It has also been shown that despite castration levels of plasma testosterone, DHT levels in the prostate itself remain at 15–40% that at baseline (Geller, 1990; Nishiyama, Hashimoto and Takahashi, 2004). These low intraprostatic levels of DHT are apparently sufficient to activate the AR and promote the expression of androgen-dependent genes (Mohler et al., 2004). Moreover, the lack of testosterone in prostate tumor cells after castration-induced androgen depletion suggests the ability of cells to bypass the standard steroidogenesis pathway in favor of a "backdoor" pathway, utilizing progesterone as a substrate for the *de novo* synthesis of DHT (Makhsida et al., 2005).

Anti-androgens such as bicalutamide have established roles, and FDA approval, for the treatment of established prostate cancer. The prolonged, sustained use of these drugs carries significant risk of side effects, such as breast pain or gynecomastia in as many as 38–39% of treated men, as well as loss of libido and muscle mass, that may restrict the broad application of the drugs to prostate cancer prevention. Nonetheless, if men at very high risk of prostate cancer development could be identified, and anti-androgenic agents could reduce this risk, the risk–benefit ratio of drug treatment might be favorable. Another approach might be to test anti-androgens in intermittent dosing schedules to ascertain whether drugs given this way might reduce prostate cancer risks with more acceptable side effects (Bruchovsky et al., 2001).

Selective estrogen receptor modulating (SERM) agents under development at a number of pharmaceutical companies, including SCH 57050 (Schering) and

LY353381 (Eli Lilly), have activity in the treatment of established prostate cancer or in prostate cancer prevention. SCH 57050 binds the estrogen receptor (ER) and inhibits ligand-independent receptor activation by polypeptide growth factors. LY353381, which binds both ER$\alpha$ and ER$\beta$, has demonstrated pre-clinical activity against the human prostate cancer cell line LNCaP propagated as a xenograft tumor in immunodeficient mice. Most SERMs will probably be clinically assessed first for therapeutic value in breast cancer and osteoporosis. Nevertheless, if selected SERMs display pre-clinical efficacy against prostate cancer models, then clinical development of these drugs not only for prostate cancer prevention, but also for treatment of advanced disease, will be rigorously pursued.

# 8
# Apoptotic-Based Molecular
# Markers of Therapeutic Response

Despite advances in clinical staging and therapy over the past few decades, current treatment and accurate staging of prostate cancer patients is not optimal. PSA and peri-operative clinical staging remain the best available predictors of disease. There are still limitations in predicting those patients with microscopic foci of advanced disease and treatment resistant cells that may lead to disease progression and ultimately death. The key morphological features diagnostic of PIN and invasive tumor cells are changes in the nuclear morphology, such as enlargement of the nucleus, profound alterations in chromatin structure, and marked nucleolar enlargement. Apoptosis, as a molecular process of genetically regulated cell death, has a critical endpoint that coincides with the goal of successful treatment of human malignancies. Since in cancer treatment the therapeutic goal is to trigger tumor-selective cell death, activation of the apoptotic pathway in prostatic tumor cells offers attractive and potentially effective therapeutic targets. As our understanding of the vital role of apoptosis in the development and growth of the prostate gland has expanded, numerous genes that encode apoptotic regulators, and which are severely impaired in prostate tumors, have been identified.

Human prostate cancer cells undergo apoptosis in response to androgen ablation, chemotherapeutic agents, and ionizing irradiation. Expression of apoptotic modulators within individual prostate tumors appears to correlate with the cancer cell's sensitivity to traditional therapeutic modalities, including radiotherapy. No strict correlation between radiation-induced apoptosis and longevity of prostate cancer patients has emerged, possibly because the ability to achieve an initial remission alone does not adequately predict long-term outcome and patient survival. Understanding the tumor-impairing effects of apoptosis-driven therapeutic modalities in prostatic cancer will lead to the identification of distinct molecular markers predictive of therapeutic response of prostatic tumors to androgen ablation, radiation therapy, chemotherapy, or combination regimens, thus enabling alternative prognostic indicators in optimizing our treatment protocols for advanced disease.

There is considerable controversy at present surrounding the nature and availability of a reliable predictor of therapeutic response and successful outcome after radiation treatment of prostate cancer patients. Identifying a common indicator of disease persistence and/or recurrence after treatment of prostate cancer with radiotherapy is of utmost importance. The American Society for Therapeutic Radiation and Oncology (ASTRO) consensus conference defines a biochemical failure as three consecutive PSA rises and backdating the time of failure to half way between the nadir and first rise with no specific nadir defined (American Society for Therapeutic Radiation and Oncology Consensus Panel, 1997). However, individual studies suggest specific cut-off values as indicative of recurrence risk. Several years ago, it was suggested that lower PSA nadirs should be used as defining disease freedom (Critz *et al.*, 1997). Evidence from a wide spectrum of studies suggests that the lower the PSA nadir level achieved and maintained, such as <0.5–1 ng/ml, the more reasonable it is to consider that there has been a successful treatment response after radiation therapy. A progressive increase in PSA over time indicates disease recurrence (Zietman *et al.*, 1994; D'Amico *et al.*, 1998; Davis *et al.*, 1999). The National Radiation Oncology Group consensus statement supports the data indicating that the nadir level of post-treatment PSA is a valuable prognostic factor, just as the pre-treatment PSA and Gleason's score are. However, it also states that the nadir value by itself is not indicative of disease status and that there is no level that guarantees success or failure of treatment (Davis *et al.*, 1999). PSA nadir by itself does appear to have some utility in selected patients, but it would be preferable to develop a way to better predict outcome prior to therapy, rather than after treatment. PSA nadir is prognostic of the time it takes to reach castration-recurrent prostate cancer and death. However, it is not known whether pre-treatment serum PSA levels can predict response to androgen deprivation therapy. These limitations of PSA have been "stigmatizing" its value as a marker of therapeutic response, and emphasize the immediate need for novel biomarkers to accurately monitor tumor aggressiveness and predict therapeutic response of prostate cancer patients to androgen deprivation therapy or radiotherapy. This is where novel biochemical or immunohistochemical predictive markers, either independent of or used in conjunction with PSA, would allow the clinician to offer alternative and/or adjuvant therapies to patients who have a high risk of recurrent or persistent disease (Sonmez *et al.*, 1995).

Development of CRPC tumors is a consequence of lack of an apoptotic response to androgen ablation, radiotherapy, and other chemotherapeutic modalities; overcoming this androgen independence is the most critical therapeutic endpoint towards improving patient survival. The use of apoptosis players as immunohistochemical markers may prove useful in the clinical staging of prostate cancer patients, and in identifying those patients who may benefit from adjuvant therapy and to predict disease-free survival (Huang *et al.*, 1998; Grossfeld, Small and Carroll, 1998; Scherr *et al.*, 1999; Szostak *et al.*, 2001). Exhaustive evaluation of key apoptosis players established that overexpressed mutant p53 and/or BCL-2 proteins are potent promoters of cell survival despite treatment with radiotherapy (reviewed by McKenzie and Kyprianou, 2006), evidence feeding a controversy surrounding their role as markers of therapeutic response in the clinical setting. Thus while the correlation of p53 with treatment failure has been challenged by studies showing no difference in cancer-specific survival and rates of p53 immunopositivity (Stattin *et al.*, 1996), others show a strong independent correlation with progression-free survival, development of metastases, and overall survival (Lee and Bernstein 1993; Grignon *et al.*, 1997; McDonnell *et al.*, 1997). Overexpression of BCL-2 and p53 proteins prolongs cell survival despite exposure to irradiation and they are useful, either independently or together, in pre-treatment biopsies for predicting response to radiotherapy for localized disease (Huang *et al.*, 1998). Further translational retrospective studies that included a larger patient population demonstrated that the apoptotic index and BCL-2 expression levels were of major significance in predicting clinical outcomes after brachytherapy of prostate cancer patients. Specifically, BCL-2 immunoreactivity was significantly higher in prostatic tumors resistant to radiotherapy compared with radiation responders (Mackey *et al.*, 1998; Szostak *et al.*, 2001). Therefore alternative immunohistochemical and/or biochemical markers, such as the BCL-2 family members, p53, and caspases (Winter *et al.*, 2001) may prove to be important independent or adjuvant predictors of resistance to current radiotherapy approaches (either external beam radiation or brachytherapy), and may open new avenues to implementation of novel therapies in combating advanced prostate cancer.

Considerable barriers to obtaining tumor tissue in large cohorts of CRPC patients at the beginning of treatment, and at relapse, preclude an effective strategy to resolve the involvement of apoptosis-driven mechanisms in the emergence of therapeutic resistance by analyzing sufficient and suitable clinical material. Rapidly accumulating knowledge of the molecular and cellular changes occurring during the transition to CRPC provides promising new windows of opportunity for the effective treatment of prostate cancer patients with metastatic disease. CRPC remains dependent on a functional AR, AR-mediated processes, and on the availability of intraprostatic intracellular androgens. CRPCs might, however, acquire different molecular mechanisms that enable them to more efficiently use intracellular androgens (AR amplification, AR overexpression, AR hypersensitivity), use alternative splice variants of the AR towards androgen-independent AR functioning, and have dramatically altered co-regulator expression (that is, changes in co-activator and co-repressor status, impacting the outcome of AR-regulated gene expression) (Dehm *et al.*, 2008; Knudsen and Scher, 2009). Driven by the intensity of such evidence, massive experimental and pre-clinical efforts focused on the characterization of CRPC have validated the concept that AR activity is regained as part of disease progression and consequently can no longer be the sole therapeutic target for advanced prostate cancer (Gregory *et al.*, 2004; Yuan and Balk, 2009; Knudsen and Scher, 2009; Zhu and Kyprianou, 2010).

Further clinical investigations with eventual therapeutic endpoints are required to implement these novel screening approaches prior to radiotherapy, in order to select prostate cancer patients with deregulation of apoptotic determining factors, who may benefit from more aggressive curative therapy.

# 9

# Role of Apoptosis in Prostate Cancer Prevention

Cancer prevention research has become a priority for several organizations globally, in areas as diverse as early detection and screening, diet and nutrition, smoking, and chemoprevention. At the cellular level, apoptosis induction has been recognized as the most potent defense against cancer initiation and metastatic spread (Reed, 1999). Primary prevention refers to the prevention of transformation and manifestation of the very first malignant cell, while tertiary prevention is used to describe the delayed progression of those cancer cells to a distant site. The ultimate chemopreventive approach entails the use of agents that quickly and effectively eliminate pre-malignant cells (before they become malignant) by inducing apoptosis, rather than promoting some degree of differentiation. Powerful experimental evidence (reviewed by Sun, Hail and Lotan, 2004) suggests that certain chemopreventive agents can induce apoptosis. Epidemiological evidence indicates that the high rates of prostate cancer morbidity and mortality in the United States might be preventable, and that a series of pharmacological approaches can be implemented to reduce the risk of prostate cancer (recently reviewed by Rittmaster, Fleshner, Thompson et al., 2009). Autopsy studies of US men dying of causes unrelated to prostate cancer suggest that the development of life-threatening prostate cancer from small neoplastic prostate lesions may take decades (Sakr et al., 1994). This slow progression of pre-clinical prostate cancer to clinical disease provides an excellent opportunity to delay the natural progression of the tumor to advanced disease. Because of the characteristic slowly progressing nature of most prostate tumors, even if we can only prevent cancers not destined to kill, simply reducing the burden of the disease can be of tremendous benefit in unnecessary treatments, their side effects, and costs. Widespread geographic variations in prostate cancer incidence and mortality exist throughout the world, with high risks of prostate cancer death characteristic of the United States and Western Europe, but not of rural Asia. Although there is a significant genetic component to prostate cancer, the environment plays an even more prominent role. Within a generation of migrating to the United States, Japanese or Chinese migrants from low-risk prostate cancer geographic regions to high-risk areas tend to adopt higher prostate cancer risks (Crawford, 2003). Moreover, case-control studies in low and high prostate cancer risk areas implicate dietary factors as modulators of prostate cancer development (Nelson, 2007). The evidence being gathered at a striking rate directly indicates that some feature of prostate cancer progression is very sensitive to environmental influences such as diet, offering great hope that some intervention might reduce prostate cancer mortality risks in the United States. Indeed, recent epidemiological observations identified a number of dietary components and micronutrients, including selenium, lycopene, and soy isoflavones (Lippman *et al.*, 2009), as candidate prostate cancer preventative agents, possibly through the induction of apoptosis among cancer cells. Clinical testing of several of these dietary components and micronutrients is currently under way, as new leads for prostate cancer prevention have been generated from investigative efforts into the molecular pathogenesis of prostate cancer.

As androgens are required for normal prostatic development, and because androgen deprivation is the gold-standard treatment for advanced prostate cancer, androgen signaling has traditionally been an attractive target for prostate cancer prevention. Inhibitors of $5\alpha$-reductase, an enzyme necessary for the conversion of testosterone to the more potent androgen dihydrotestosterone, have reached pivotal clinical trials for prostate cancer prevention. In addition, new insights into the molecular pathogenesis of prostate cancer hint that chronic or recurrent prostate inflammation may contribute to the development of the disease. A variety of antioxidants and anti-inflammatory drugs, likely to be capable of attenuating pro-carcinogenic genome damage from reactive oxygen and nitrogen species, are also under current development for prostate cancer prevention.

Various inflammatory stimuli, including sexually transmitted infections (STIs), dietary factors, exposure to estrogen, and urine reflux, have been implicated as potential environmental triggers of prostatic inflammation (De Marzo *et al.*, 2007). Young men with STIs are reported to be more likely to have an increase in serum PSA than men without STIs, suggesting that such infections can damage the prostate epithelium, leading to the leakage of PSA into the bloodstream, even at a young age. However, though a number of STIs have been correlated with prostate cancer incidence, a causal relationship has not been established. A novel retrovirus, xenotropic murine leukemia virus related virus (XMRV), was found in prostate tumors among men with genetic defects in RNASEL, evidence suggesting that an interaction between genetic susceptibility and environmental exposure might trigger and/or propagate prostatic inflammation (Dong *et al.*, 2007).

There is also epidemiologically-based evidence revealing an increased risk of prostate cancer among men with a history of prostatitis. A systematic review of 11 epidemiological studies found that men with a history of prostatitis had a 1.6% higher risk (95% CI: 1.0, 2.4) of prostate cancer (Dennis, Lynch and Tomer, 2002), although this could be partly due to a detection bias, as patients with symptomatic prostatitis are more likely to visit a urologist and thus are more likely to be screened for prostate cancer. Reflux of urine has been implicated in promoting prostatic inflammation, presumably by damaging the prostatic epithelium and by stimulating the production of inflammatory cytokines via activation of NALP3 inflammasomes by the prostate (Sfanos *et al.*, 2009).

With regard to diet, meat intake has long been associated with increased prostate cancer risk, a risk that may partially be explained by cooking practices. Cooking meats at high temperatures, especially charbroiling fatty meats, can result in the generation of heterocyclic amines (HCAs), such as PhIP and others, as well as in polycyclic aromatic hydrocarbons (PAHs). Both can induce mutations, and cause cell and tissue damage (Felton *et al.*, 2007). Experimental studies in rats have shown that PhIP is a potent carcinogen through its ability to trigger severe epithelial cell damage, inflammation, PIA, and cancer selectively in the ventral lobe of the prostate gland (Nakai, Nelson and De Marzo, 2007). Diets rich in fruits and vegetables, particularly tomatoes (which contain antioxidants such as lycopene), and cruciferous vegetables (which contain chemoprotective isothiocyanates such as sulforaphane) have been associated with a reduced risk of prostate cancer (Cohen, Kristal and Stanford, 2000). Moreover, dietary n-3 polyunsaturated fatty acids enhance hormone ablation therapy in androgen-dependant prostate cancer (McEntee *et al.*, 2008). Chemopreventive intervention with green tea extract has been shown in animal models of prostate tumorigenesis to proceed via apoptosis induction, thereby inhibiting tumor progression to metastasis (Gupta *et al.*, 2001).

## 9.1 Aspirin and Non-Aspirin Nonsteroidal Anti-Inflammatory Drugs (NSAIDs)

Aspirin and non-aspirin non-steroidal anti-inflammatory drugs (NSAIDs), which act to attenuate inflammation by inhibiting cyclooxygenases (COXs), have received widespread attention as potential chemopreventive agents for a number of different cancers (Cuzick *et al.*, 2009). Observational studies revealed an inverse association between these agents and total cancer incidence (Jacobs *et al.*, 2007), and with cancer-related mortality (Bardia *et al.*, 2007), particularly for breast and prostate cancer (Harris *et al.*, 2005). Although aspirin and the non-aspirin NSAIDs broadly inhibit both COX isoforms (cyclooxygenase-1 (COX-1) and cyclooxygenase-2 (COX-2)), newer drugs such as celecoxib and rofecoxib selectively block COX-2, providing anti-inflammatory activity but less of the adverse effects associated with COX-1 blockade such as gastrointestinal bleeding. There is abundant pre-clinical, clinical, and epidemiological data to support the potential use of the COX-2 inhibitor, celecoxib, in prostate cancer chemoprevention (Sooriakumaran *et al.*, 2007). Knowledge of the molecular mechanisms underlying NSAID-mediated apoptosis has been available since the late 1990s (Chan *et al.*, 1998b). Experimental studies documented that selective COX-2 inhibitors suppressed the growth of human prostate cancer cells *in vitro* and *in vivo* through induction of apoptosis, and inhibition of cell proliferation and angiogenesis (Hsu *et al.*, 2000; Liu *et al.*, 2000; Kamijo *et al.*, 2001). Such actions were readily translated in the clinical context as prospective cohort studies have shown a significant association between the use of non-selective COX-2 inhibitors and reduced risk of prostate cancer (Garcia Rodriguez and González-Pérez, 2004; Jacobs *et al.*, 2005; Harris *et al.*, 2005). However, pre-clinical studies using selective and non-selective COX inhibitors for prostate cancer prevention and treatment have generated confusing results. Thus although COX-2-selective inhibitors have been recognized as apoptosis inducers, able to reduce the growth of various human prostate cancer xenografts *in vivo*, whether this effect is a direct consequence of COX-2 inhibition in prostate cancer cells, or in host cells in the prostate tumor microenvironment is not clearly established (Narayanan *et al.*, 2006; Narayanan *et al.*, 2009). Since COX-2 is not produced in human prostate cancer cells (Zha *et al.*, 2001), this concept is clearly significant in understanding the cellular targeting by the COX-2 inhibitors. Several previous studies demonstrating COX-2 expression in human prostate cancer cells used defective immunohistochemistry methods. The absence of COX-2 expression in human prostate cancer cases may be attributable to epigenetic silencing of COX-2 (Yegnasubramanian *et al.*, 2004). Another confounding phenomenon is that selective and non-selective COX inhibitors appear also to promote inactivation of BCL-2, inactivation of AKT, reduction of NF-κB p65, down-regulation of androgen receptor activity, up-regulation of the p75 NTR tumor suppressor, and increase in detoxification enzyme GSTP1, all of which could contribute to potential tumor-inhibitory effects. Population-based studies of aspirin intake have hinted at a protective effect on prostate cancer risk. In a systematic meta-analysis involving 12 studies published prior to 2003, Mahmud and colleagues (Mahmud *et al.*, 2006), found a significant inverse association between aspirin use and prostate cancer, and this was confirmed in subsequent studies (Jacobs *et al.*, 2005; Platz *et al.*, 2005). Observation of the fascinating inverse association between aspirin use and prostate cancer indicates that it appears stronger for higher frequency use, for greater duration of use, among younger men, and for advanced prostate cancer.

The association between non-aspirin NSAIDs and prostate cancer in population studies is more varied, as a review found an inverse association between non-aspirin NSAIDs with prostate cancer, while other studies continued to report mixed results (Platz *et al.*, 2005; Dasgupta *et al.*, 2006). An apparently critical and unresolved issue that might add clarity to the population studies concerns the fraction of ingested aspirin or of the different non-aspirin NSAIDs appearing in prostate tissues and/or the lumens of prostate glands (location of many of the inflammatory cells). Carefully designed pharmacological studies, using drugs such as aspirin and desipramine as model agents, have suggested that factors such as hydrophobicity, pH-$pK_a$ partitioning, and metabolism might affect the propensity of drugs to enter the prostate gland and seminal vesicles (Cao *et al.*, 2008). In order to prevent the onset of prostatic carcinogenesis, drugs that are incapable of accumulating in the prostate at sufficiently high concentrations to attenuate inflammation are unlikely to generate a biological effect and great clinical benefit.

A series of large clinical trials evaluating the effect of aspirin on colorectal cancer incidence, and on total cancer incidence among women, has yielded varying degrees of successful outcomes (Cook *et al.*, 2005). However, there has been no randomized clinical trial testing the effect of aspirin on prostate cancer incidence among male populations. Three independent clinical trials (Pruthi *et al.*, 2006; Smith *et al.*, 2006; Carles *et al.*, 2007) focused on celecoxib as a therapeutic modality, in combination with taxanes or radiotherapy in prostate cancer patients, but it is challenging to derive meaningful conclusions due to early interruption of these trials because of safety issues associated with the intake of this class of drug. Furthermore, a planned large randomized prostate cancer prevention trial of another COX-2 selective inhibitor (rofecoxib) was abandoned when it was removed from the market, owing to concerns about increased cardiovascular risks, despite much disappointment from patients, clinicians, and scientists. It was not a totally "lost cause" however, since informative data have been generated in pursuit of a randomized clinical trial of celecoxib, given at a dose of 400 mg twice daily to a small number of men with prostate cancer (n=64) in a four to six week period before radical prostatectomy. In this trial, the selective COX-2 inhibitor showed very little, if any, effect on prostate tissues removed at surgery. Specifically, the drug had no effect on the appearance or extent of PIA lesions, indicating that epithelial damage and associated inflammation was not attenuated (Bardia *et al.*, 2009).

Flurbiprofen, historically an admixture of S-flurbiprofen, a cyclooxygenase inhibitor, and R-flurbiprofen, an anti-inflammatory agent that does not inhibit cyclooxygenases but may alter COX-2 mRNA levels (Bjerre and LeLorier, 2001), has been used for many years as an NSAID. R-flurbiprofen alone, which has exhibited intriguing pre-clinical efficacy in the TRAMP mouse prostate cancer model, has entered phase I–II human clinical testing against established prostate cancer. Exisulind (sulindac sulfone), originally considered as an NSAID, can also affect cGMP-dependent protein kinases. Exisulind is being subjected to clinical testing in a randomized, placebo-controlled trial for treatment of established prostate cancer, manifest as a rising serum without overt metastatic disease after primary prostate cancer treatment. Additional clinical studies, using exisulind to treat early-stage prostate cancer before radical prostatectomy, promise to deliver high-impact results. 5-lipoxygenases have been found to serve as critical mediators of cell fate decisions in prostatic carcinoma cells *in vitro* (Dale *et al.*, 2006; Browning and Martin, 2007; Bonovas, Filioussi and Sitaras, 2008). In the future, 5-lipoxygenase inhibitors may be developed as drugs for prostate cancer treatment and prevention.

Considered together, the basic, pre-clinical, epidemiological and clinical trial evidence supports further study of the protective effects of aspirin and non-aspirin NSAIDs against prostate cancer development. Identifying subpopulations of patients most likely to respond to NSAIDs based on biomarker profile, and utilizing NSAIDs as targeted therapies, is most likely to optimize success. Given the potential gastrointestinal adverse effects of NSAIDs, the recently reported combination with proton pump inhibitors also appears an attractive option (Cuzick *et al.*, 2009). On the other hand, given the serious adverse cardiovascular effects, there should be little enthusiasm to focus such efforts on selective COX-2 inhibitors.

## 9.2 Statins

Statins inhibit the enzyme 3-hydroxy-3-methylglutaryl-coenzyme A (HMG-CoA) reductase involved in the cholesterol synthesis pathway. The drugs can be classified as hydrophilic (atorvastatin, fluvastatin, pravastatin, and rosuvastatin) versus lipophilic (lovastatin and simvastatin), based on solubility in water or lipids, respectively. The predominant clinical use of statins has been to lower cholesterol in order to reduce cardiovascular disease risk; however recent studies point to the attractive possibilities of the chemopreventive potential of these drugs for a number of human malignancies, including prostate cancer (Farwell *et al.*, 2008). At the molecular level, and at relatively high concentrations, statins are capable of inducing apoptosis and triggering G1 cell cycle arrest, ultimately causing a significant suppression of growth of human prostate cancer cells. This action is thought to reflect inactivation of Ras and Rho GTPases involved in the regulation of the cell cycle (Hogue, Chen and Xu, 2008). Statins have also been reported to have anti-inflammatory properties, probably mediated by the modulation of endothelial nitric oxide synthase and peroxisome proliferator-activated receptors (PPARs).

Epidemiological studies assessing the relationship between statins and prostate cancer revealed mixed results regarding the effects on prostate cancer incidence. However, the results are subject to detection bias, not providing a control for PSA testing. Thus, while statin users tend to be more health conscious and more likely to undergo PSA screening more than their counterparts, one may argue that initiation of statin medication can result in a decline in PSA levels (Hamilton *et al.*, 2008). Interestingly, the studies that controlled for PSA testing have consistently demonstrated an inverse association between statin use and advanced prostate cancer (Platz *et al.*, 2006; Flick *et al.*, 2007; Jacobs *et al.*, 2007; Murtola *et al.*, 2007; Boudreau *et al.*, 2008). The reduction has been reported to occur with both lipophilic and lipophobic statins, to be more prominent after being taken for five years or more, and to provide a synergistic benefit when combined with NSAID use.

Four recent systematic reviews of randomized controlled trials have evaluated the efficacy of statins in the prevention of prostate cancer, and none found any association between statins and prostate cancer (Bjerre and LeLorier, 2001; Baigent *et al.*, 2005; Dale *et al.*, 2006; Browning and Martin, 2007). However, the systematic reviews did not assess the efficacy of statins on advanced prostate cancers, and the median follow-up of the trials was less than five years. One has to also consider that prostate cancer was a secondary outcome in most of the clinical trials and thus subject to detection and reporting bias. A recent systematic review by Bonovas (Bonovas, Filioussi and Sitaras, 2008) combined data from 13 observational (six cohort and seven case control studies) and six randomized control trials, and found no evidence between statin use and total prostate cancer incidence among either RCTs or the observational studies. However, statin use was inversely related to advanced prostate cancer among observational studies.

Despite the heterogeneously gloomy evidence so far on the efficacy of statins for prostate cancer chemoprevention, there is considerable promise associated with the use of these agents. The development of biomarkers predicting response to statins, and facilitating their use as targeted therapy would ensure rational utilization of these therapies as chemopreventive agents, a strategy that would be comparable with the use of NSAIDs. A combination of NSAIDs and statins as combined chemopreventive agents also appears to be an attractive option.

## 9.3 Antioxidants

Chronic inflammation is associated with oxidative stress and free radical production, which can damage genomic DNA and enhance the development of prostate carcinogenesis, particularly in the presence of androgens (Ames and Wakimoto, 2002). Men with deficiencies in antioxidant enzymes, such as polymorphisms in OGG1 (which encodes an enzyme necessary for damage repair to oxidized DNA), or in SOD2, (which encodes manganese superoxide dismutase, a mitochondrial enzyme that protects cells from oxidative damage), have been found to have a higher risk of prostate cancer. While various antioxidants have been proposed to have chemopreventive properties, vitamin E (various tocopherols) and selenium have received great attention for the prevention of prostate cancer. α-tocopherol (α-vitamin E) is the major vitamin isoform that is preferentially bound by tocopherol transfer protein (TTP) to transport from liver to serum in humans. Besides its antioxidant function, α-vitamin E and its analogs can attenuate lipid peroxidation, induce cell cycle arrest via up-regulation of p27, induce apoptosis (Gunawardena, Murray and Meikle, 2000), inhibit AR signaling pathways, and suppress prostate cancer growth via inhibition of the phosphoinositide 3-kinase pathway (Gunawardena, Murray and Meikle, 2000; Zhang *et al.*, 2002a; Zu and Ip, 2003; Ni *et al.*, 2005). It was also reported that expression of the tocopherol-associated protein that facilitates vitamin E absorption by the cell (TAP) was significantly down-regulated in PIN and in the malignant prostate compared to the normal prostate (Ni *et al.*, 2005). Selenium is postulated to decrease oxidative damage to DNA, to promote apoptosis, and to inhibit prostate cancer cell proliferation (Waters *et al.*, 2003). The chemopreventive benefits of selenium may perhaps be mediated at the cellular level through its function as a component of small selenium-containing metabolites, or potentially because of its role in controlling the activity of selenium-containing proteins or selenoproteins. The latter concept has gained validation by *in vivo* evidence indicating that selenoprotein-deficient mice exhibited accelerated development of lesions associated with prostate cancer progression, directly implicating selenoproteins in the risk and development of prostate cancer (Diwadkar-Navsariwala *et al.*, 2006). The link between selenium and prostate cancer risk has been supported by thorough multiple observational prospective studies which documented an inverse relationship between levels of selenium, as well as of tocopherols, and prostate cancer risk (Yoshizawa *et al.*, 1998; Brooks *et al.*, 2001; Helzlsouer, *et al.*, 2000). The inverse association has been observed predominantly in patients with advanced prostate cancers, usually becomes apparent after at least five years of supplementation, and is higher among men with baseline PSA levels greater than 4 ng/dl (Li *et al.*, 2004; Kirsh *et al.*, 2006).

In the 1990s, two large randomized controlled chemoprevention trials reported that alpha-tocopherol and selenium supplementation reduced the incidence of prostate cancer (Clark *et al.*, 1996; The Alpha-Tocopherol Beta Carotene Cancer Prevention Study Group, 1994), generating excitement and hope among male populations and investigators eager to pursue this prevention potential further. The original hope was reinforced by results from two other larger trials that followed (Herceberg *et al.*, 2004; Li *et al.*, 2004) and in a systematic review of these trials (Bardia *et al.*, 2008). However, two recent large randomized controlled trials, Selenium and Vitamin E Cancer Prevention

Trial (SELECT) and Physicians' Health Study II failed to confirm these findings (Lippman *et al.*, 2009; Gaziano *et al.*, 2009). SELECT, the largest chemopreventive trial conducted to date, reported that selenium alone, vitamin E alone, or their combination had no effect on prostate cancer incidence (Lippman *et al.*, 2009). Similarly, in the Physician Health II study, neither vitamin E nor vitamin C had an effect on prostate cancer incidence (Gaziano *et al.*, 2009). The emerging questions surrounding the failure of the two recent trials to detect the protective effects of antioxidant supplementation against prostate cancer development seen in earlier studies must be considered. First, the NPC trial, which found that selenium supplementation reduced the incidence of prostate cancer, targeted a population thought to be nutritionally deficient in selenium. In this trial, as was predicted by epidemiological studies, the beneficial effects of selenium in reducing prostate cancer were predominantly restricted to men with low plasma selenium concentrations (Duffield-Lillico *et al.*, 2003). Similarly, the ATBC trial, which found that vitamin E supplementation reduced prostate cancer incidence, targeted male smokers, known to have lower vitamin E levels (Clark *et al.*, 1996). Hence the possibility that the negative results from SELECT were due to administration of supplements to a nutritionally replete (rather than deficient) population. In fact, a recent follow-up of the General Population Nutrition Intervention Trial found that people who received factor D (50 μg selenium, 30 mg vitamin E, and 15 mg beta-carotene) had lower overall mortality ($P=0.009$), particularly among those men younger than 55 years (Qiao *et al.*, 2009). Moreover, one should consider the distinct possibility that SELECT may have used the incorrect supplement preparations. For instance, γ-tocopherol may provide a greater benefit than the α-tocopherol used for SELECT (Helzlsouer *et al.*, 2000). In the SELECT study, the serum levels of γ-tocopherol actually declined with α-tocopherol supplementation. The selenium used for SELECT, l-selenomethionine, was not the selenium used in the NPC trial, which gave high-selenium yeast. Finally, the data from several large cohort studies have suggested that a reduction in prostate cancer that might be associated with vitamin E and selenium supplementation would be predominantly seen for advanced prostate cancers, and would be most evident after at least five years of supplementation (Li *et al.*, 2004; Kirsh *et al.*, 2006). In SELECT, the majority of prostate cancers detected were low-grade, early-stage tumors. It is to be hoped that future analyses of SELECT may detect men with either low antioxidant levels or specific genetic vulnerability, who might benefit from targeted use of supplements. In the meantime, the advice that should be given to diet-conscious populations is based on our one consistent finding: the consumption of vegetables rich in antioxidant micronutrients is associated with diminished prostate cancer risk. This suggests that antioxidant intake might better be accomplished by dietary modulation than by supplement pills. The pathogenesis of human prostate cancer, featuring the emergence of a chronic or recurrent inflammatory state in the prostate in response to epithelial damage, probably occurs over many decades. Rational prostate cancer prevention strategies must involve a combination of primary prevention against exposures, such as to HCAs and PAHs in charbroiled meats, which inflict prostate epithelial damage, and of therapeutic approaches for interference with the procarcinogenic inflammatory processes that drive cancer development.

## 9.4 5α-Reductase Inhibitors

Testosterone, the major circulating androgen in men, is converted to the major intracellular androgen, DHT, by steroid 5α-reductase isoenzymes, designated as type 1 and type 2. The increase of the conversion rate of testosterone into DHT by an increase of the isoenzyme 5α-reductase type 1 (5AR1) through SRD5A1 gene up-regulation is well documented (Titus *et al.*, 2005; Stanbrough *et al.*, 2006). Identification of individuals with congenital deficiency of 5AR has served as a naturally occurring model ultimately leading to the development of a pharmacologic inhibitor of 5AR. Results from the Prostate Cancer Prevention Trial (PCPT) have sparked interest in the potential roles of the 5α-reductase inhibitors in the management and prevention of prostate cancer (Thompson *et al.*, 2003).

Finasteride, marketed as Proscar for the treatment of benign prostatic hyperplasia (BPH) and as Propecia for alopecia, is a selective inhibitor of the type II 5α-reductase that catalyzes the conversion of testosterone to dihydrotestosterone in the prostate (Geller, 1990). Finasteride is an orally active, competitive inhibitor of the nicotanimide adenine dinucleotide phosphate, reduced from the type 2 isozyme of the NADPH-dependent 5AR enzyme. The drug has a fairly attractive side effect profile and a dramatic effect in suppressing serum DHT to baseline levels in men. Several large-scale clinical trials have established the efficacy of finasteride in the medical management of BPH (McConnell *et al.*, 2003). More recently, it was shown that finasteride can cause significant reductions in the serum PSA and/or delayed rises in the serum PSA in a number of men with established prostate cancer (Thompson *et al.*, 2006b), and indeed this was the very first drug found to reduce the risk of prostate cancer. Two large-scale clinical trials investigated the possible role of 5AR inhibitors in the prevention of prostate cancer: The Prostate Cancer Prevention Trial (PCPT) (Thompson *et al.*, 2003) and the Reduction by Dutasteride of Prostate Cancer Events (REDUCE) (Andriole *et al.*, 2004).

In the Prostate Cancer Prevention Trial (PCPT), finasteride was effective in men at both low and high risk for prostate cancer (Thompson *et al.*, 2006a). Neither age, PSA level at enrollment, family history of prostate cancer, race nor ethnicity affected the drug's ability to prevent the disease (Thompson *et al.*, 2004). Despite the fact that men taking finasteride had fewer prostate tumors overall, they had a greater proportion of high-grade prostate cancer (Thompson *et al.*, 2003). Of those men on finasteride, 6.4% (280 out of 4,368) had high-grade tumors, while 5.1% of men on placebo (237 out of 4,692) had high-grade cancers. Despite a strong relative risk reduction of approximately 25% in the diagnosis of prostate cancer in the PCPT study, there were several issues obscuring a widespread acceptance of finasteride (a safe, FDA-approved agent) as a preventative strategy for prostate cancer. Interestingly enough, having a low PSA level did not correlate with the development of aggressive tumors, as some of the men in both groups of the trial had high-grade disease despite their PSA level (Thompson *et al.*, 2007). Consequently, the PCTP was met with profound skepticism by some clinicians, investigators and patients, fuelled by concerns that finasteride had selected for, and accelerated the growth of, high-grade tumors (Scardino, 2003). Serious concerns were raised among the urologic community as to the detection bias and interpretation of the results from the PCPT (Goodman *et al.*, 2006; Cohen *et al.*, 2007). The consideration here is that although a larger percentage of men taking finasteride had tumors that appeared to be pathologically more aggressive, the aggressive behavior of the tumor at the molecular and cellular level is not clear (Lucia *et al.*, 2007). A long-term follow-up is required for these men to determine whether a high-grade tumor in patients taking finasteride correlates with aggressive disease biologically.

The recently reported results of the expanded REDUCE trial by Andriole and colleagues on the effect of dutasteride in reducing the risk of prostate cancer (Andriole *et al.*, 2010) are timely and immensely significant. The study represented a four-year, international, placebo-controlled, randomized trial that evaluated the effect of dutasteride in reducing the presence of cancer, as assessed on random biopsies performed after two and four years of treatment. Significantly, there was a 23% relative reduction in incident prostate cancer detected on biopsy over the course of four years (a rate of 25.1% in the placebo group versus 19.9% in the dutasteride group). The skeptics argue that the findings most likely represent suppression of the growth of existing tumors (presumably via apoptosis induction), rather than prevention of cancer. This recent evidence raises critical questions about whether finding a decrease in cancer detected in random biopsies is a meaningful endpoint, and whether there was a reduction in the incidence of tumors with lethal potential. Thus, neither dutasteride nor finasteride significantly reduced the risk of prostate cancer among men who were followed closely and who underwent a biopsy because of an elevated PSA level (corrected for the effect of the drug), or an abnormal digital rectal examination. What the recently reported study established is that the dual 5α-reductase inhibitor dutasteride reduced the incidence of prostate cancer detected on biopsy among men who had an increased risk for prostate cancer (Andriole *et al.*, 2010). This reduction was observed mainly among men who had tumors with Gleason scores of 5 or 6. These findings must be considered seriously when designing a setting that would be used for optimization and refinement of future chemoprevention trials.

# 10

# Summary and Future Directions

The focus on cell proliferation as the core aberrant characteristic of cancer has historically meant that almost all anti-cancer drug efforts have been developed as anti-proliferative, anti-mitotic interventions for the treatment of human cancer. These drugs have without question proven clinically useful in shrinking tumors, and although this increases survival or delays disease recurrence, cure of established epithelial cancers is uncommon. Total eradication of all cancer cells never occurs; tumor regrowth proceeds from residual cells, and whatever benefits are gained from the use of anti-mitotic drugs are often at the cost of considerable toxicity, because cell proliferation is so intrinsic to the viability and function of normal cells (Okada and Mak, 2004). This is a heavy price to pay for the organism, as sacrifice of the normal cells to eliminate the tumor mass does not provide a cure. With the identification of apoptosis as an equally important partner as cell proliferation in the growth equation and in maintaining organ homeostasis, the recognition of the loss of this process (interpreted as reduced capacity of cells to die) during tumorigenesis directed efforts at pharmacologic development towards the therapeutic value in its "reinstating" apoptosis. The concept assumes more significant dimensions in the case of prostate cancer, because prostate tumor growth dynamics have a special characteristic feature, distinct from other solid tumors. It is this concept which has provided the basis for this monograph. The molecular basis of prostate cancer includes inheritable and somatic genetic changes (tumor suppressor genes, loss of heterozygosity, gene targets and regions of chromosomal gain, CpG island promoter methylation, invasion and metastasis suppressor genes, telomere shortening, and genetic instability) (Kote-Jarai *et al.*, 2008). Changed gene expression (such as proliferation-related genes, changes in apoptosis and AR genes) have potential as biomarkers and therapeutic targets in prostate cancer. Since prostate cancer development and progression results from roadblocks in the apoptotic signaling pathways, rather than uncontrolled cell proliferation, apoptosis-based strategies to kill tumor cells without affecting normal cells involve direct induction of pro-apoptotic molecules, modulation of either anti-apoptotic or survival proteins, or restoration of tumor-suppressor gene functions (Zhang *et al.*, 2002b; McCarty, 2004; McKenzie and Kyprianou, 2006). Apoptosis mechanistically proceeds through one of two signaling cascades, both of which converge on activating the executioner caspases, caspase-3 and caspase-7. In the intrinsic pathway, mitochondrial outer membrane permeabilization (MOMP) drives initiator caspase activation and apoptosis, by releasing pro-apoptotic proteins from the mitochondrial intermembrane space. Upon release from the mitochondria, cytochrome c binds to APAF-1, inducing its conformational change, apopto-some formation, and ultimately activation of caspase-9 which in turn activates capase-3 and capase-7 in a cascade fashion (Kim *et al.*, 2008).

As the first decade of the 21st century ends, advanced, CRPC, and meta-static prostate cancer continues to represent a significant healthcare burden. There is active research into the specific biochemical steps in the apoptosis mechanisms that become dysfunctional during the initiation of prostate cancer and progression to metastatic disease (primarily to the bone), and which can be potentially re-instated in the tumor cells but not the normal cells within the tumor microenvironment (Gleave *et al.*, 1991; Henshall *et al.*, 2001). The need is urgent, as the disease claims the life of a large number of men. The previously named hormone-refractory or androgen-independent prostate cancer is now officially classified as castration-resistant prostate cancer (CRPC), and, as the lethal stage of the malignancy, is responsible for 30,000 deaths among American men each year. The widely popularized belief that currently there is no gold-standard chemotherapy regimen for castration-resistant prostate cancer that has consistently shown objective tumor regression or survival benefit must be revisited as we enter the era of personalized medicine. The "standard rules", generalized by several generations of clinicians for an entire patient population with striking tumor heterogeneity, are seen to be simplistic. The complexity of the genetic mechanisms leading to apoptosis and providing the foundation for apoptosis-driven therapies suggests that the acquisition of a castration-resistant phenotype by a prostate cancer cell is likely to have been achieved by a multitude of signaling pathways, some of which are engaged in a dynamic, highly complex cross-talk. Moreover, the prostatic epithelium is composed of AR-negative cells within the basal compartment, AR-positive secretory cells, and AR-negative neuroendocrine cells (Arnold and Isaacs, 2002). Whereas the secretory epithelial cells are truly androgen-dependent and undergo apoptosis in response to androgen withdrawal, the basal cell compartment is androgen-insensitive, not really requiring androgens for its survival. Pioneering studies from the autopsy series at the University of Michigan provided valuable lessons and the long-sought information that androgen-independent prostate cancer represents an extensively heterogeneous disease (Shah *et al.*, 2004). Reinstating apoptosis-signaling pathways in prostate tumors is of paramount importance but of immense complexity, since resistance to apoptosis at the biochemical level in cells is responsible for conferring therapeutic failure in patients, and anoikis resistance is a major contributor to the metastatic spread of tumor cells (Sakamoto and Kyprianou, 2010). However, it is clear from the above discussion of existing clinical and translational evidence that this effort is not a simple one-directional exercise at the molecular level that would lead to a well-defined single agent, administration of which would eradicate an entire tumor cell population in a group of patients with CRPC. In addition to targeting the core components of the apoptotic machinery (BCL-2, IAPS, APAF-1 and caspases) (Gleave *et al.*, 1999; Nikitina *et al.*, 2005; Kim *et al.*, 2008), cellular opportunities exist to indirectly impact apoptosis outcomes by modulating inputs into cell death pathways through protein kinases, protein phosphatases, nuclear transcription factors (Jiao *et al.*, 2006), and cell membrane receptors for growth factor signaling pathways (such as EGFR) (Lyons *et al.*, 2001; Bjornsti and Houghton, 2004a). The interplay platform is richly populated by the primers of death, and detailed dissection of their role has potentially important consequences for the development of novel anti-tumor therapies aimed at restoring apoptosis, reversing anoikis resistance, inhibiting EMT, and impairing metastasis (Reginato *et al.*, 2003). Prostate tumors can become resistant to apoptosis during progression to metastasis via multiple molecular mechanisms, including:

- expression of mitochondrial gatekeeper proteins of the BCL-2 family;
- overexpression of inhibitors of apoptosis (Amantana *et al.*, 2004);
- loss of expression of caspases (primarily caspase-8 of the death receptor apoptosis signaling pathway, caspase-9 of the mitochondrial pathway, and caspase-3 (the downstream apoptosis final executioner) (Winter *et al.*, 2001); and
- overexpression of critical focal adhesion complex players sufficient to confer anoikis resistance, such as talin 1 (Sakamoto *et al.*, 2010).

Moreover, the hypoxia response signaling system also emerges as a therapeutic target for prostate cancer via its strong contribution to the tumor vascularity (Cvetkovic *et al.*, 2001; Anastasiadis *et al.*, 2003). The cross-talk between anoikis and apoptosis converges at the key activation point of CD95/FAS as it connects RIP, a kinase that shuttles between CD95/FAS-mediated death and FAK-mediated survival pathways (Kischkel *et al.*, 1995; Sprick *et al.*, 2000; Horbinski, Mojesky and Kyprianou, 2010).

The early clinical success of two experimental drugs that target persistent AR pathway signaling in CRPC (abiraterone and MDV31000) validates the hypothesis that restored AR function is a major driver of CRPC progression. Abiterone is a potent steroidal inhibitor of the adrenal and testicular enzyme CYP17, and blocks residual adrenal production in castrate patients already treated with androgen-ablative agents such as leuprolide (Attard *et al.*, 2008). MDV31000 is a second-generation competitive AR antagonist discovered to have a strong anti-tumor action in CRPC animal models (Tran *et al.*, 2009). Both abiraterone and MDV31000 are expected to be approved for the treatment of CRPC patients in coming years, pending the results of ongoing randomized phase III trials in chemotherapy-refractory CRPC patients where survival is an endpoint. While there is substantial excitement about the clinical activity of both of these agents, the medical community is quick to recognize the limitations associated with their modest therapeutic benefit.

Recent breakthroughs in the development of novel AR-antagonist strategies led to phase I clinical trials with the potential to improve the efficacy of AR targeting and consequently the therapeutic outcome in patients with CRPC (Tran *et al.*, 2009). Development of effective combinatorial therapeutic approaches might be required to successfully eliminate CRPC and prevent recurrence. Therapeutic benefit for a subset of patients with locally advanced disease stems from a combination of radiation therapy with androgen ablation towards improved response (Harris *et al.*, 2009). Since the publication of two randomized trials that showed significant survival and palliative benefits for men with prostate cancer treated with docetaxel, this drug has become the treatment of choice for patients with metastatic CRPC (De Dosso and Berthold, 2008). Despite the slow acceptance of docetaxel as a standard effective treatment for advanced disease due to modest survival benefit (2–3 months), the clinical knowledge gathered so far regarding its potent targeting in CRPC calls for further understanding of the drug's mechanisms of action in order to augment its therapeutic efficacy. Indeed, there is rapidly growing evidence in experimental and clinical studies that taxanes can target prostate tumors by alternative routes besides mitosis disruption, and this may assist the development of combination strategies for treating CRPC. The evidence that docetaxel counteracts the pro-survival effects of BCL-2 gene expression by inducing phosphorylation and inhibition of the survival protein product BCL-2 is of critical significance (Haldar, Jena and Croce, 1995; Oliver *et al.*, 2005) in understanding the mechanisms underlying therapeutic resistance to taxanes in individual patients. Moreover, our recent investigations of the effect of microtubule-targeting drugs on AR signaling in prostate tumors revealed that microtubule-targeting agents, such as paclitaxel impair AR nuclear transport and activity in clinical specimens from prostate cancer patients (Zhu and Kyprianou, 2010). The concept that concurrent administration of AR-antagonizing strategies with docetaxel may impede the side effects of the chemotherapeutic agent has been validated in mouse models of prostate cancer (Hess-Wilson *et al.*, 2006). Significantly, clinical support for this is emerging from clinical trials that indicate an improved response to docetaxel in the presence of androgens (Hussain *et al.*, 2005; Rathkopf *et al.*, 2008). These findings collectively provide a foundation for considering the AR input into the development of combinatorial approaches with existing drugs (such as the microtubule-targeting agents), or novel compounds (Gan *et al.*, 2009; Zhu *et al.*, 2010b). Careful exploitation of the molecular mechanisms that are likely to drive the optimization of the combinatorial modality is essential, as suppression of AR-activity might not always be leading to apoptosis induction of the prostate tumor epithelial cells, depending on the tumor microenvironment and the hypoxic state of the tumor (Nicholson *et al.*, 2004).

The technology-driven future offers exotic therapeutic options for exploiting existing drugs used for the treatment of other cancers or other diseases, or novel molecular targets for which the signaling pathways have been exhaustively analyzed. An added feature of selective apoptosis-signaling targeting in castration-resistant prostate cancer cells is the prospect of "reversed pharmacology", which enables identification of molecular targets of efficacious pharmacological agents with an unknown mechanism of action. The two clinically available quinazoline-based $\alpha_1$-adrenoceptor antagonists, doxazosin and terazosin, are currently in clinical use as anti-hypertensive agents and as a standard medical therapy for benign prostatic hyperplasia (BPH) treatment (Kyprianou, 2003). The therapeutic benefit of these drugs has been attributed to their ability to induce smooth muscle relaxation and reducing vascular wall dynamics. In recent years, cancer researchers worldwide documented that drugs with a safety profile, already in clinical use for different conditions — such as the $\alpha_1$-adrenoceptor antagonists, anti-inflammatory drugs, and statins — can exert potent anti-growth effects via induction of epithelial cell apoptosis (Tahmatzopoulos and Kyprianou, 2004; Harris *et al.*, 2007). The newly identified growth-suppressing properties of the quinazolines have three distinct characteristics that distinguish them from existing cytotoxic drugs used for prostate cancer treatment:

1) the quinazoline effect activates apoptosis exclusively without affecting cell proliferation of prostate tumor epithelial cells;
2) this action by the piperazinyl quinazolines proceeds via novel mechanisms that are independent of the $\alpha_1$-adrenoceptor (a sulfonamide based $\alpha_1$-antagonist, tamsulosin fails to elicit an apoptotic response in prostate cancer cells); and
3) the drugs suppress prostate tissue vascularity by inhibiting cell adhesion and invasion via anoikis.

Homelessness is one of a multitude of environmental stresses that can befall a cell (Horbinski, Mojesky and Kyprianou, 2010). When cancer cells undergo a shortage of oxygen, nutrients, or hormones, they metastasize, invade local tissue, enter the lymph and blood stream, and establish a colony at a distant site from the primary tumor. This strategy of cell survival promotes the dissemination of the cancer cell population, but is often fatal to the patient. The events involved in tumor cell progression towards increasing metastatic potential are largely mediated by the integrins, which upon engagement with components of the ECM, reorganize to form adhesion complexes (Timpl *et al.*, 2003). The current understanding of anoikis-governing mechanisms (apoptosis consequential to insufficient cell-matrix interactions) led to the definition of a role for this mode of death intrinsic to prostate tumor angiogenesis and metastasis. Reversed programming of the anoikis-style execution by the cells, once they are detached and homeless, renders them free and invasive, and ultimately successful in their metastatic spread to the bone (Simpson, Anyiwe and Schimmer, 2008). Recent evidence from pre-clinical studies indicates that targeting specific signaling effectors of anoikis (such as SRC, talin, and AKT) may prove promising as a therapeutic approach for impairing prostate cancer invasion and metastasis (Nam *et al.*, 2005; Garrison *et al.*, 2007; Park *et al.*, 2008). Furthermore, the promise of a potential therapeutic impact by a combination of an apoptosis-based and anoikis-targeted drug is very appealing at the mechanistic, as well as at the translational level. Integrin $\beta_1$ deregulation in response to anoikis "re-sensitization" by emerging drugs is potentially linked to alterations in the focal adhesion complex, and key components of the actin microfilaments that determine cell motility (Goel *et al.*, 2008; Sakamoto and Kyprianou, 2010). The goal is to maximize the therapeutic value of the novel agents to overcome anoikis resistance (by AKT activation), and to suppress prostate tumor invasion and metastasis in order to treat advanced prostate cancer.

The functional outcome of EMT in the progression of prostate cancer to castration-resistant disease is complex, given the uncertainty surrounding the contribution of the androgen axis to prostate cancer metastasis. Indeed, skepticism remains about whether EMT truly occurs during prostate cancer progression, and whether it plays an indispensable role in metastasis. The impact of androgen suppression on metastatic dissemination of prostate cancer cells is still a subject of debate, and there is a notion that androgen deprivation therapy may down-regulate AR in prostate tumor-circulating cells. The significance of EMT during clinical cancer progression has also been vigorously debated, due to the lack of convincing evidence of EMT in clinical specimens on primary and metastatic prostate cancer. Many argue that in the absence of histological identification of EMT intermediates, it is challenging to distinguish individual mesenchymal cells derived from epithelial tumor cells after EMT, from stromal cells or from other tumor-associated fibroblasts (Tlsty and Hein, 2001). To resolve this apparent contradiction, a mesenchymal–epithelial transition (MET) process in the metastatic sites has been postulated to explain histopathological similarities in primary tumors and metastatic lesions (Tsuj, Ibaragi and Hu, 2009). Situated at the interface of tumor and stroma signals, EMT reflects an intricate counterbalance between internal growth pressure (by an expanding tumor) and the free edge of the migrating tumor periphery (Savagner, 2001;

Thompson, Newgreen and Tarin, 2005). TGF-β takes a critically acclaimed role in the regulation of bone metastasis of prostate cancer cells, and has also been shown to be a potent regulator of EMT causally linked to bone metastasis of prostate cancer cells. Cadherin switching has been strongly established during prostate cancer progression (Tomita *et al.*, 2000) and the association of the EMT phenotype, as predicted by E-cadherin loss (Onder *et al.*, 2008), with progression to metastasis and therapeutic resistance, is likely to have a strong clinical impact in prostate cancer patients. Detection of metastatic lesions at an early stage of the disease or during treatment should increase disease-free survival rates. This calls for the pursuit of signaling pathways that will allow "smart" therapeutic targeting of anoikis-resistant tumor circulating cells undergoing EMT at the initiation of metastasis (Barrallo-Gimeno and Nieto, 2005; Thompson, Newgreen and Tarin, 2005). This research has begun, as therapeutic promise was recently reported in a proteosome inhibitor impairing EMT in metastatic prostate cancer by suppressing *Snail* and inducing *RKIP* (Raf kinase inhibitor protein), despite the skepticism surrounding the role of EMT in malignant development (Tarin, Thompson and Newgreen, 2005; Hugo *et al.*, 2007).

Therapeutic exploitation of the ESR apoptotic pathways is reaching a new zenith. Researchers are developing the theory that the SERCA pump serves as the surrogate therapeutic target in prostate cancer, as proteolysis in PSA-producing sites will hydrolyze a prodrug such as thapsigargin (a well-known pharmacological inducer of ER stress) into an active agent with extended bioavailability in the tumor sites (Treiman, Caspersen and Christensen, 1988; Denmeade and Isaacs, 2005). A functional interrogation of ER-specific death signals within the context of apoptotic machinery specific to CRPC prostate tumor cells, and possibly their surrounding microenvironment, leads to the BCL-2 family members as ultimate apoptosis controllers of cellular death and survival decisions (Rutkowski and Kaufman, 2004). A closer look at the regulation of calcium homeostasis by the ER provides a direct link to the BCL-2 function, since upon BCL-2 phosphorylation, $Ca^{2+}$ release from the ER into the cytoplasmic pools is increased, with a secondary increase in mitochondrial $Ca^{2+}$ uptake, thus enabling an intriguing mechanism by which post-translational modification of the apoptosis suppressor BCL-2 blocks its survival activity (Bassik *et al.*, 2004). However, targeting these genes in prostate cancer cells by engaging ESR pathways does not secure a selective therapeutic impact relative to other apoptotic outcomes that do not engage the ER as the death platform. Targeting the ESR/UPR pathways may be the ultimate mechanism to treat prostate cancer, and indeed this is inviting and full of promise, but much needs to be learned first, and fast (Hampton, 2002).

Regardless of the cellular process linking tumor behavior to its suicidal death, the striking dependence of prostate tumor epithelial cells on apoptosis pathways emphasizes the huge impact of apoptosis research in the clinical setting of human disease on two fronts:

a) the elucidation of cancer-relevant pathways, proteins, and biochemical exchanges, opening the way for application of "smart" therapy based on well-defined biological mechanisms; and

b) the identification of tumor biomarkers that will allow proper diagnosis, disease staging/categorization, and clinically relevant prognostic information.

Apparently seductive combinations that must be considered include abrogation of growth factor signaling, or transcriptional regulatory pathways that selectively contribute to ligand-independent AR activity. Valuable insights into the context of the tumor microenvironment in prostate cancer growth and metastasis (Pienta and Bradley, 2006) would ultimately be provided by careful and methodical dissection of the mechanisms of action of growth factor signaling up-regulated in prostate tumors after androgen ablation. These include the survival mechanisms leading to tumor cell proliferation, enhanced vascularity, deregulated inflammation, and resistance to chemotherapy (Che and Grignon, 2002; Vicentini *et al.*, 2003, Gao *et al.*, 2003; Farhana *et al.*, 2004; Festuccia *et al.*, 2005), as well as the mechanisms of AR deregulation in "metastatic" prostate cancer (Lu *et al.*, 2004). Inflammation has also been implicated in indirectly negating the inhibitory effects of AR antagonists and converting them into agonists (Zhu *et al.*, 2006). Combinatorial approaches are being formulated, capitalizing on the anti-tumor apoptosis-driven action of known therapeutic modalities (such as taxanes or ionizing irradiation) being enhanced by administration of angiogenesis-targeting agents (angiostatin) in CRPC treatment (Galaup *et al.*, 2003; Zhu *et al.*, 2010a). Intense efforts are currently invested in powerful proteomic and mass spectrometry methods to discern how membrane-dependent pathways dictate aggressive tumor behavior. Identification of a predictive marker on a molecular signature before or during pre-clinical anti-tumor drug development will enable the selection of patients who are likely to respond and exhibit therapeutic benefit (Ding *et al.*, 2011).

The functional interplay between the androgen axis and lead cytokine signaling pathways, such as FGF, IGF-1, EGF, VEGF, and TNF may impact the outcomes of apoptosis-driven therapeutic targeting in prostate tumors (Reynolds and Kyprianou, 2006; Zhu and Kyprianou, 2008). Rigorous targeting of these growth factor signaling pathways and their transcriptional effectors that contribute to ligand-independent AR activity could lead to successful elimination of the prostate tumors regardless of their AR status (Culig and Bartch, 2006; Léotoing *et al.*, 2007; Knudsen and Scher, 2009). Moreover, long-term exposure of prostate cancer cells to TNF-$\alpha$ leads to hypersensitivity to androgens and anti-androgen depletion (Harada *et al.*, 2001). Thus, while antagonists of these signaling mechanisms could in some cases act in concert with androgen ablation and AR antagonists towards suppression of prostate tumor growth and progression to CRPC, in other cases such a combination might not be therapeutically beneficial. The functional promiscuity of growth factors can render the pathways they elicit from survival mechanisms into apoptosis effects, under conditions of AR depletion (Zhu and Kyprianou, 2010). Hence, designing combination approaches targeting frequently overlapping signaling effectors may provide the ultimately effective modality for treating patients with advanced metastatic CRPC, using a "personalized medicine" approach. Specifically, dissection of the interaction of TGF-β and AR can lead to new therapeutic options based on abrogation of the cross-talk between TGF-β, its receptors, and AR (Zhu and Kyprianou, 2008). Significantly enough, the TGF-β promoter was shown to contain three distal and three proximal androgen-response elements (AREs), that physically interact with the DNA-binding domain of AR (as reviewed by Knudsen and Scher, 2009). This evidence identifies molecular targeting of the androgen-TGF-β signaling cross-talk as an attractive therapeutic approach in prostate cancer (Zhu and Kyprianou, 2005). Ongoing studies suggest combining the targeting of TGF-β signaling with novel therapeutic compounds as AR antagonists.

As the links between stromal collagen density, ECM integrity, and tumor progression solidify, the underlying molecular mechanisms are slowly unraveling. The ECM-crosslinking enzyme lysyl oxidase (LOX) is overexpressed in many tumors including prostate cancer. This up-regulation is associated with poor prognosis; LOX is also involved in recruiting inflammatory stromal cells that contribute to tumor progression (Erler *et al.*, 2009) pointing out its therapeutic targeting value by use of anti-inflammatory drugs. Anti-inflammatory drugs and antioxidants may be capable of attenuating prostate cancer pathogenesis, particularly advanced cancers, after prolonged use. For these agents to be effective, intracellular accumulation in prostate tissue is a prior requirement, and they must be able to inhibit inflammation and/or oxidant damage; pharmacodynamic studies, therefore, should be developed, in addition to efficacy studies. While aspirin and statins appear promising as chemopreventive agents mechanistically driving the apoptosis signaling pathways, development of biomarkers that predict response to these therapies, and facilitate their use as targeted therapies alone or in combination, would ensure rational and effective utilization of these therapies as chemopreventive agents. An example of such an apoptosis-inducing agent is psoralidin, a herbal molecule and an active ingredient of the plant *Psoralea corylifolia*. This has hepatoprotective properties and osteoblastic-proliferating activity. The AKT-survival targeting action of this molecule was recently shown to be translated into a potent anti-tumor effect against CRPC *in vitro* and *in vivo* (Kumar *et al.*, 2009).

Chemoprevention of prostate cancer is a challenging, laudable, and high-impact public health initiative. Since the tumorigenic process inherently selects against apoptosis to initiate and promote the malignant phenotype towards metastasis (anoikis resistance), targeting apoptosis-signaling pathways in pre-malignant prostate epithelial cells is clearly effective in preventing prostate cancer progression. The striking new insights into the molecular and genetic mechanisms of apoptosis regulation during prostatic carcinogenesis constitute new opportunities for the use of drugs, as well as dietary components, to prevent the onset and progression of prostate cancer effectively. Serious exploitation of such opportunities will be driven by chemoprevention drug discovery and development, requiring coordinated research efforts within large pharmaceutical companies and small biotechnology companies, in academia, and at the National Cancer Institute. Pivotal clinical trials, if designed to ascertain the effects of new drugs on prostate cancer survival in the general population, will likely take many years, increasing commercial development risks while decreasing exclusive marketing revenues. New initiatives at the FDA transfer some chemoprevention clinical trial costs to the health insurance and health maintenance industries, and changes in patent laws may ultimately reduce commercial risks associated with the development of prostate cancer chemotherapy drugs, making their discovery more attractive to pharmaceutical and biotechnology companies. Over the past decade, remarkable progress has been made in elucidating the molecular events associated with prostate cancer development, and new reagents and technologies associated with the National Institutes of Health Human Genome Project and the National Cancer Institute Cancer Genome Anatomy Project have the potential to accelerate the pace of discovery even further. The recent results of the REDUCE trial, in combination with ongoing PCTP analyses of the massive dataset gathered, may justify targeting androgen metabolic pathways after all, to prevent the onset of this cancer that is so characteristically dependent on androgens for its development. Modeling data from the PCTP trial, and analysis of prostate biopsy specimens from men treated with dutasteride for four months before surgery, support the finding that the major biological action of dutasteride at the cellular level is the shrinkage of prostate tumors via apoptosis of androgen-sensitive prostate epithelial cells. For the immediate future, prostate cancer chemoprevention drug development research needs validated surrogate and strategic clinical trial endpoints, defined high-risk clinical trial cohorts, and new drugs for clinical testing.

The emerging opportunities are driven by powerful novel bioinformatics tools and databases, knowledge of the functional roles of proteins and protein modifications in the cellular context (Ding *et al.*, 2011), new imaging tools that allow high-resolution detection of cellular and tissue events such as gene fusions (despite the cellular heterogeneity that characterizes prostatic tumors) (Mosquera *et al.*, 2008), as well as prediction of differential sensitivity of tumors to apoptosis-based therapeutic approaches in widespread model systems that allow a precise functional assessment of death signaling pathways. Nanotechnology will drive therapeutic outcomes (Maynard and Puis, 2007) by securing drug delivery within tumors (Sugahara *et al.*, 2009), becoming "armour" phenomenally suited for the fight against cancer. It is up to us to seize the opportunities towards finding the cure.

# Bibliography

Aaronson, D. S., Muller, M., Neves, S. R., *et al.* (2007). An androgen-IL-6-Stat3 autocrine loop re-routes EGF signal in prostate cancer cells, *Mol. Cell. Endocrinol.* **270(1–2)**: 50–56.

Abdollahi, A., Hahnfeldt, P., Maercker, C., *et al.* (2004). Endostatin's antiangiogenic signaling network, *Mol. Cell* **13(5)**: 649–663.

Abdulghani, J., Gu, L., Dagvadorj, A., *et al.* (2008). Stat3 promotes metastatic progression of prostate cancer, *Am. J. Pathol.* **172(6)**: 1717–1728.

Abreu-Martin, M. T., Chari, A., Palladino, A. A., *et al.* (1999). Mitogen-activated protein kinase kinase kinase 1 activates androgen receptor-dependent transcription and apoptosis in prostate cancer, *Mol. Cell Biol.* **19(7)**: 5143–5154.

Acloque, H., Adams, M. S., Fishwick, K., *et al.* (2009). Epithelial-mesenchymal transitions: the importance of changing cell state in development and disease. *J. Clin. Invest.* **119(6)**: 1438–1449.

Adams, J. (2004). The proteasome: a suitable antineoplastic target, *Nat. Rev. Cancer* **4**: 349–360.

Adams, J., Palombella, V. J., Sausville, E. A., *et al.* (1999). Proteasome inhibitors: a novel class of potent and effective antitumor agents, *Cancer Res.* **59**: 2615–2622.

Affar, E. B., Luke, M. P-S., Gay, F., *et al.* (2006). Targeted ablation of Par-4 reveals a cell type–specific susceptibility to apoptosis-inducing agents, *Cancer Res.* **66**: 3456.

Agarwal, A., Munoz-Najar, U., Klueh, U., *et al.* (2004). N-acetyl-cysteine promotes angiostatin production and vascular collapse in an orthotopic model of breast cancer, *Am. J. Pathol.* **164(5)**: 1683–1696.

Agoulnik, I. U., and Weigel, N. L. (2008). Androgen receptor coactivators and prostate cancer, *Adv. Exp. Med. Biol.* **617**: 245–255.

Akhurst, R. J. (2002). TGF-β antagonists: Why suppress a tumor suppressor? *J. Clin. Invest.* **109(12)**: 1533–1536.

Akhurst, R. J., and Derynck, R. (2001). TGF-β signaling in cancer – a double-edged sword, *Trends Cell Biol.* **11(11)**: S44–S51.

Albig, A. R., and Schiemann, W. P. (2004). Fibulin-5 antagonizes vascular endothelial growth factor (VEGF) signaling and angiogenic sprouting by endothelial cells, *DNA Cell Biol.* **23**(6): 367–379.

Algire, G. H., and Chalkley, H. W. (1945). The vascular supply of mammary gland carcinomas. [Accessed 10 November 2010] http://tgmouse.compmed.ucdavis.edu/preprint/Kenney/Algire%20and%20Chalkley.pdf

Allinen, M., Beroukhim, R., Cai, L., *et al.* (2004). Molecular characterization of the tumor microenvironment in breast cancer, *Cancer Cell* **6**(1): 17–32.

Altieri, D. C. (2003). Validating survivin as a cancer therapeutic target, *Nat. Rev. Cancer* **3**(1): 46–54.

Amantana, A., London, C. A., Iversen, P. L., *et al.* (2004). X-linked inhibitor of apoptosis protein inhibition induces apoptosis and enhances chemotherapy sensitivity in human prostate cancer cells, *Mol. Cancer Ther.* **3**(6): 699–707.

Amato, R. J., and Sarao, H. (2006). A Phase I Study of Paclitaxel/Doxorubicin/Thalidomide in Patients with Androgen-Independent Prostate Cancer. *Clin. Genitourin. Canc.* **4**(4): 281–286.

American Society for Therapeutic Radiation and Oncology Consensus Panel (1997): Consensus statement: guidelines for PSA following radiation therapy, *Int. J. Rad. Oncol. Biol. Phys.* **37**: 1035.

Ames, B. N., and Wakimoto, P. (2002). Are vitamin and mineral deficiencies a major cancer risk? *Nat. Rev. Cancer* **2**: 694–704.

Anastasiadis, A. G., Bemis, D. L., Stisser, B. C., *et al.* (2003). Tumor cell hypoxia and the hypoxia-response signaling system as a target for prostate cancer therapy, *Curr. Drug Targets* **4**(3): 191–196.

Anderson, R. J., Mawji, N. R., Wang, J., *et al.* (2010). Regression of castrate-recurrent prostate cancer by a small-molecule inhibitor of the amino-terminus domain of the androgen receptor, *Cancer Cell* **17**: 535–546.

Andriole, G. L., Bostwick, D. G., Brawley, O., *et al.* (2004). Chemoprevention of prostate cancer in men at high risk: rationale and design of the reduction by dutasteride of prostate cancer events (REDUCE) trial, *J. Urol.* **172**: 1314–1317.

Andriole, G. L., Bostwick, D. G., Brawley, O. W., *et al.* (2010). Effect of Dutasteride on the risk of prostate cancer, *N. Engl. J. Med.* **362**: 1192–1202.

Ao, M., Williams K., Bhowmick, N. A., *et al.* (2006). Transforming growth factor-beta promotes invasion in tumorigenic but not in nontumorigenic human prostatic epithelial cells, *Cancer Res.* **66**(16): 8007–8016.

Aoudjit, F., and Vuori, K. (2001). Matrix attachment regulates FAS-induced apoptosis in endothelial cells: a role for c-flip and implications for anoikis, *J. Cell Biol.* **152**(3): 633–643.

Aoyama, M., Asai, K., Shishikura, T., *et al.* (2001). Human neuroblastomas with unfavorable biologies express high levels of brain-derived neurotrophic factor mRNA and a variety of its variants, *Cancer Lett.* **164**(1): 51–60.

Apakama, I., Robinson, M. C., Walter N. M., *et al.* (1996). BCL-2 overexpression combined with p53 protein accumulation correlates with hormone-refractory prostate cancer, *Br. J. Cancer* **74**: 1258–1262.

Apel, A., Zentgraf, H., Buchler, M. W., *et al.* (2009). Autophagy – A double-edged sword in oncology, *Int. J. Cancer* **125**: 991–995.

Arap, W., Haedicke, W., Bernasconi, M., *et al.* (2002). Targeting the prostate for destruction through a vascular address, *Proc. Natl. Acad. Sci. USA*, **99**(3): 1527–1531.

Armstrong, A. J., Netto, G. J., Rudek, M. A., *et al.* (2010). A pharmacodynamic study of rapamycin in men with intermediate- to high-risk localized prostate cancer, *Clin. Cancer Res.* **16**(11): 3057–3066.

Arnold, J. T., and Isaacs, J. T. (2002). Mechanisms involved in the progression of androgen-independent prostate cancers: it is not only the cancer cell's fault, *Endocr. Relat. Canc.* **9**: 61–73.

Arsura, M., Panta, G. R., Bilyeu, J. D., *et al.* (2003). Transient activation of NF-kappaB through a TAK1/IKK kinase pathway by TGF-beta1 inhibits AP-1/SMAD signaling and apoptosis: implications in liver tumor formation, *Oncogene* **22**(3): 412–425.

Ashkenazi, A. (2002). Targeting death and decoy receptors of the tumour-necrosis factor superfamily, *Nat. Rev. Cancer* **2**(6): 420–430.

Attard, G., Reid, A. H., Yap, T. A., *et al.* (2008). Phase 1 clinical trial of a selective inhibitor of CYP17, abiraterone acaetate, confirms that castration-resistant prostate cancer commonly remains hormone-driven, *J. Clin. Oncol.* **26**: 4563–4571.

Attwell, S., Roskelley, C., and Dedhar, S. (2000). The integrin-linked kinase (ILK) suppresses anoikis, *Oncogene* **19**: 3811–3815.

Ayala, G., Thompson, T., Yang, G., *et al.* (2004). High levels of phosphorylated form of Akt-1 in prostate cancer and non-neoplastic prostate tissues are strong predictors of biochemical recurrence, *Clin. Cancer Res.* **10**(19): 6572–6578.

Bachman, K.E., and Park, B. H. (2005). Duel nature of TGF-[beta] signaling: tumor suppressor vs. tumor promoter. *Curr. Opin. Oncol.* **17**(1): 49–54.

Baigent, C., Keech, A., Kearney, P. M., *et al.* (2005). Efficacy and safety of cholesterol-lowering treatment: prospective meta-analysis of data from 90,056 participants in 14 randomised trials of statins, *Lancet* **366**: 1267–1278.

Balk, S. P., and Knudsen, K. E. (2008). AR, the cell cycle, and prostate cancer. *Nucl. Recept. Signal* **6**: 1–12.

Balkwill, F., and Coussens, L. M. (2004). An inflammatory link, *Nature* **431**: 405–406.

Bandyopadhyay, A., Wang, L., Lopez-Casillas, F., *et al.* (2004). Systemic administration of a soluble betaglycan suppresses tumor growth, angiogenesis, and matrix metalloproteinase-9 expression in a human xenograft model of prostate cancer. *Prostate* **63**(1): 81–90.

Banham, A. H., Boddy, J., Launchbury, R., *et al.* (2007). Expression of the forkhead transcription factor FOXP1 is associated both with hypoxia inducible factors (HIFs) and the androgen receptor in prostate cancer but is not directly regulated by androgens or hypoxia. *Prostate* **67**(10): 1091–1098.

Barbero, S., Mielgo, A., Torres, V., *et al.* (2009). Caspase-8 association with the focal adhesion complex promotes tumor cell migration and metastasis. *Cancer Res.* **69**: 3755.

Bardia, A., Ebbert, J. O., Vierkant, R. A., *et al.* (2007). Association of aspirin and nonaspirin nonsteroidal anti-inflammatory drugs with cancer incidence and mortality, *J. Natl. Cancer Inst.* **99**: 881–889.

Bardia, A., Platz, E. A., Yegnasubramanian, S., *et al.* (2009). Anti-inflammatory drugs, antioxidants and prostate cancer prevention, *Curr. Opin. Pharmacol.* **9**: 419–426.

Bardia, A., Tleyjeh, I. M., Cerhan, J. R., *et al.* (2008). Efficacy of antioxidant supplementation in reducing primary cancer incidence and mortality: systematic review and meta-analysis, *Mayo Clin. Proc.* **83**: 23–34.

Barrallo-Gimeno, A., and Nieto, M. A. (2005). The Snail genes as inducers of cell movement and survival: implications in development and cancer, *Development* **132**(14): 3151–3161.

Barrett, T., Kobayashi, H., Brechbiel, M., *et al.* (2006). Macromolecular MRI contrast agents for imaging tumor angiogenesis. *Eur. J. Radiol.* **60**(3): 353–366.

Baserga, R., Peruzzi, F., and Reiss, K. (2003). The IGF-1 receptor in cancer biology, *Int. J. Cancer* **107**(6): 873–877.

Bassik, M. C., Scorrano, L., Oakes, S. A., *et al.* (2004). Phosphorylation of BCL-2 regulates ER Ca2+ homeostasis and apoptosis, *EMBO J.* **23**(5): 1207–1216.

Basu, A., and Haldar, S. (1998). Microtubule-damaging drugs triggered bcl-2 phosphorylation: requirement of phosphorylation on both serine-70 and serine-87 residues of bcl-2 protein, *Int. J. Oncol.* **13**: 659–664.

Bauer, B., Jenny, M., Fresser, F., *et al.* (2003). AKT1/PKBalpha is recruited to lipid rafts and activated downstream of PKC isotypes in CD3-induced T cell signaling, *FEBS Lett.* **541**(1–3): 155–162.

Beck, C., Schreiber, H., and Rowley, D. A. (2001). Role of TGF-β in immune-evasion of cancer, *Microsc. Res. Techniq.* **52**: 387–395.

Beer, T. M., Garzotto, M., Lowe, B. A., *et al.* (2004). Phase I study of weekly mitoxantrone and docetaxel before prostatectomy in patients with high-risk localized prostate cancer, *Clin. Cancer Res.* **10**: 1306–1311.

Beer, T. M., Pierce, W. C., Lowe, B. A., *et al.* (2001). Phase II study of weekly docetaxel in symptomatic androgen-independent prostate cancer, *Ann. Oncol.* **12**: 1273–1279.

Bello-DeCampo, D., and Tindall, D. J. (2003). TGF-β/Smad signaling in prostate cancer, *Curr. Drug Targets* **4**(3): 197–207.

Bender, F. C., Reymond, M. A., Bron, C., *et al.* (2000). Caveolin-1 levels are down-regulated in human colon tumors, and ectopic expression of caveolin-1 in colon carcinoma cell lines reduces cell tumorigenicity, *Cancer Res.* **60**(20): 5870–5878.

Benning, C. M., and Kyprianou, N. (2002). Quinazoline-derived α1 adrenoceptor antagonists induce prostate cancer cell apoptosis via an α1-adrenoceptor independent action, *Cancer Res.* **62**: 597–602.

Berezovskaya, O., Schimmer, A. D., Glinski, A. B., *et al.* (2005). Increased expression of apoptosis inhibitor protein XIAP contributes to anoikis resistance of circulating human prostate cancer metastasis precursor cells, *Cancer Res.* **65**: 2378–2386.

Bergers, G., and Benjamin, L. E. (2003). Tumorigenesis and the angiogenic switch, *Nat. Rev. Cancer* **3**(6): 401–410.

Bergers, G., and Hanahan, D. (2008). Models of resistance to anti-angiogenic therapy. *Nat. Rev. Cancer* **8**(8): 592–603.

Berges, R. R., Vukanovic, J., Epstein, J. I., *et al.* (1995). Implication of cell kinetic changes during the progression of human prostatic cancer, *Clin. Cancer Res.* **1**(5): 473.

Berrier, A. L., Martinez, R., Bokoch, G. M., *et al.* (2002). The integrin beta tail is required and sufficient to regulate adhesion signaling to Rac1, *J. Cell Sci.* **115**(22): 4285–4291.

Berry, S. J., Coffey, D. S., Walsh, P. C., *et al.* (1984). The development of human benign prostatic hyperplasia with age, *J. Urol.* **132**(3): 474.

Berry, S. J., and Isaacs, J. T. (1984). Comparative aspects of prostatic growth and androgen metabolism with aging in the dog versus the rat, *Endocrinology* 114(2): 511.

Berry, W., Dakhil, S., Gregurich, M. A., *et al.* (2001). Phase II trial of single-agent weekly docetaxel in hormone-refractory, symptomatic, metastatic carcinoma of the prostate, *Semin. Oncol.* 28: 8–15.

Bertolotti, A., Zhang, Y., Hendershot, L. M., *et al.* (2000). Dynamic interaction of BiP and ER stress transducers in the unfolded-protein response, *Nat. Cell Biol.* 2(6): 326–332.

Bhowmick, N. A., Chytil, A., Plieth, D., *et al.* (2004). TGF-beta signaling in fibroblasts modulates the oncogenic potential of adjacent epithelia, *Science* 303: 848–851.

Bijnsdorp, I. V., Peters, G. J., Temmink, O. H., *et al.* (2010). Differential activation of cell death and autophagy results in an increased cytotoxic potential for trifluorothymidine compared to 5-fluorouracil in colon cancer cells, *Int. J. Cancer* 126: 2457–2468.

Biswas, S., Criswell, T. L., Wang, S. E., and Arteaga, C. L. (2006). Inhibition of transforming growth factor – β signaling in human cancer: targeting a tumor suppressor network as a therapeutic strategy, *Clin. Cancer Res.* 12: 4142.

Biswas, S. C., and Greene, L. A. (2002). Nerve growth factor (NGF) Down regulates the BCL-2 homology 3 (BH3) domain-only protein BIM and suppresses its proapoptotic activity by phosphorylation, *J. Biol. Chem.* 277: 49511–49516.

Bix, G., and Iozzo, R. V. (2005). Matrix revolutions: "tails" of basement-membrane components with angiostatic functions, *Trends Cell Biol.* 15(1): 52–60.

Bjerre, L. M., and LeLorier, J. (2001). Do statins cause cancer? A meta-analysis of large randomized clinical trials, *Am. J. Med.* 110: 716–723.

Bjornsti, M-A., and Houghton, P. J. (2004a). The mTOR pathway: a target for cancer therapy, *Nat. Rev. Cancer* 4: 335–348.

Bjornsti, M-A., and Houghton, P. J. (2004b). Lost in translation: dysregulation of cap-dependent translation and cancer, *Cancer Cell* 5(6): 519–523.

Blackledge, G. (2003). Growth factor receptor tyrosine kinase inhibitors; clinical development and potential for prostate cancer therapy, *J. Urol.* 170: S77–S83.

Blobe, G. C., Schiemann, W. P., and Lodish, H. F. (2000). Role of transforming growth factor β in human disease, *N. Engl. J. Med.* 342: 1350–1358.

Boccon-Gibod, L., Hammerer, P., Madersbacher, S., *et al.* (2007). The role of intermittent androgen deprivation in prostate cancer, *BJU Int.* 100(4): L738–743.

Boddy, J. L., Fox, S. B., Han, C., *et al.* (2005). The androgen receptor is significantly associated with vascular endothelial growth factor and hypoxia sensing via hypoxia-inducible factors HIF-1a, HIF-2a, and the prolyl hydroxylases in human prostate cancer, *Clin. Cancer Res.* 11(21): 7658–7663.

Bodmer, J., Holler, N., Reynard, S., *et al.* (2000). TRAIL receptor-2 signals apoptosis through FADD and caspase-8, *Nat. Cell Biol.* 2: 241–243.

Boehm, T., Folkman, J., Browder, T., *et al.* (1997). Antiangiogenic therapy of experimental cancer does not induce acquired drug resistance, *Nature* 390(6658): 404–407.

Bogdanos, J., Karamanolakis, D., Tenta, R., *et al.* (2003). Endocrine/paracrine/autocrine survival factor activity of bone microenvironment participates in the development of androgen ablation and chemotherapy refractoriness of prostate cancer metastasis in skeleton, *Endocr. Relat. Cancer* 10(2): 279–289.

Bolden, J. E., Peart, M. J., and Johnstone, B. W. (2006). Anticancer activities of histone deacylase inhibitors, *Nat. Rev. Drug Discov.* **5**: 769–784.

Bonaccorsi, L., Carloni, V., Muratori, M., *et al.* (2004a). EGF receptor (EGFR) signaling promoting invasion is disrupted in androgen-sensitive prostate cancer cells by an interaction between EGFR and androgen receptor (AR), *Int. J. Cancer* **112**: 78–86.

Bonaccorsi, L., Marchiani, S., Muratori, M., *et al.* (2004b). Gefitinib ('IRESSA', ZD1839) inhibits EGF-induced invasion in prostate cancer cells by suppressing PI3 K/AKT activation, *J. Cancer Res. Clin. Oncol.* **130(10)**: 604–614.

Bonaccorsi, L., Nosi, D., Muratori, M., *et al.* (2007). Altered endocytosis of epidermal growth factor receptor in androgen receptor positive prostate cancer cell lines, *J. Mol. Endocrinol.* **38(1–2)**: 51–66.

Bonham, M. J., Galkin, A., Montgomery, B., *et al.* (2002). Effects of the herbal extract PC-SPES on microtubule dynamics and paclitaxel-mediated prostate tumor growth inhibition, *J. Natl. Cancer Inst.* **94**: 1641–1647.

Bonovas, S., Filioussi, K., and Sitaras, N. M. (2008). Statin use and the risk of prostate cancer: a meta-analysis of 6 randomized clinical trials and 13 observational studies, *Int. J. Cancer* **123**: 899–904.

Borgstrom, P., Bourdon, M. A., Hillan, K. J., *et al.* (1998). Neutralizing anti-vascular endothelial growth factor antibody completely inhibits angiogenesis and growth of human prostate carcinoma micro tumors *in vivo*, *Prostate* **35(1)**: 1–10.

Bostwick, D. G., and Brawer, M. K. (1987). Prostatic-intraepithelial neoplasia and early invasion of prostate cancer, *Cancer* **59**: 788–794.

Boudreau, D. M., Yu, O., Buist, D. S. M., *et al.* (2008). Statin use and prostate cancer risk in a large population-based setting, *Cancer Cause. Control* **19(7)**: 767–774.

Bowen, C., Voeller, H. J., Kikly, K., *et al.* (1999). Synthesis of procaspases-3 and -7 during apoptosis in prostate cancer cells, *Cell Death Differ.* **6(5)**: 394–401.

Boyce, M., Bryant, K. F., Jousse, C., *et al.* (2005). A selective inhibitor of eIF2$\alpha$ dephosphorylation protects cells from ER stress, *Science* **307(5711)**: 935–939.

Brabletz, T., Jung, A., Reu, S., *et al.* (2001). Variable $\beta$-catenin expression in colorectal cancers indicates tumor progression driven by the tumor environment, *PNAS* **98(18)**: 10356–10361.

Brasel, J. A., Coffey, D. S., and Williams-Ashman, H. G. (1968). Androgen-induced changes in the DNA polymerase activity of coagulating glands of castrated rats, *Med. Exp. Int. J. Exp. Med.* **18(4)**: 321.

Breckenridge, D. G., Germain, M., Mathai, J. P., *et al.* (2003). Regulation of apoptosis by endoplasmic reticulum pathways, *Oncogene* **22(53)**: 8608–8618.

Breier, G., Blum, S., Peli, J., *et al.* (2002). Transforming growth factor-beta and Ras regulate the VEGF/VEGF-receptor system during tumor angiogenesis, *Int. J. Cancer* **97**: 142–148.

Brodie, B. C. (1834). "Lectures on Diseases of the Urinary Organs". 2$^{nd}$ edition, 1835. Longman, Ress, Orme, Brown, Green and Longman; Paternoster-Row.

Brooks, J. D., Metter, E. J., Chan, D. W., *et al.* (2001). Plasma selenium level before diagnosis and the risk of prostate cancer development, *J. Urol.* **166**: 2034–2038.

Brooks, P. C., Montgomery, A. M. P., Rosenfeld, M., *et al.* (1994). Integrin $\alpha$v$\beta$3 antagonists promote tumor regression by inducing apoptosis of angiogenic blood vessels, *Cell* **79(7)**: 1157–1164.

Browning, D. R., and Martin, R. M. (2007). Statins and risk of cancer: a systematic review and meta-analysis, *Int. J. Cancer* **120**: 833–843.

Bruchovsky, N., Goldenberg, S. L., Rennie, P. S., *et al.* (1995). Theoretical considerations and initial clinical results of intermittent hormone treatment of patients with advanced prostatic carcinoma, *Urologe A* **34**(5): 389–392.

Bruchovsky, N., Goldenberg, S. L., Mawji, N. R., *et al.* (2001). Evolving aspects of intermittent androgen blockage for prostate cancer: diagnosis and treatment of early tumor progression and maintenance of remission, *Andrology in the 21st Century, Proceedings of the VIIth International Congress of Andrology*, pp. 609–623.

Bruchovsky, N., and Wilson, D. (1968). The conversion of testosterone to 5-alpha-androstan-17-beta-ol-3-one by rat prostate *in vivo* and *in vitro, J. Biol. Chem.* **243**: 2012–2021.

Bruckheimer, E. M., and Kyprianou, N. (2000). Apoptosis in prostate carcinogenesis: a growth regulator and a therapeutic target, *Cell Tissue Res.* **301**: 153–162.

Bruckheimer, E. M., and Kyprianou, N. (2001). Dihydrotestosterone enhances transforming growth factor-beta-induced apoptosis in hormone-sensitive prostate cancer cells, *Endocrinology* **142**(6): 2419–2426.

Bruckheimer, E. M., and Kyprianou, N. (2002). BCL-2 antagonizes the combined apoptotic effect of transforming growth factor-beta and dihydro-testosterone in prostate cancer cells, *Prostate* **53**(2): 133–142.

Buijs, J. T., Henriquez, N. V., van Overveld, P. G. M., *et al.* (2007). TGF-beta and BMP 7 interactions in tumour progression and bone metastasis, *Clin. Exp. Metastasis* **24**(8): 609–617.

Burchardt, M., Burchardt, T., Shabsigh, A., *et al.* (2001). Reduction of wild type p53 function confers a hormone resistant phenotype on LNCaP prostate cancer cells, *Prostate* **48**(4): 225–230.

Burfeind, P., Chernicky, C. L., Rininsland, F., *et al.* (1996). Antisense RNA to the type I insulin-like growth factor receptor suppresses tumor growth and prevents invasion by rat prostate cancer cells *in vivo, Proc. Natl. Acad. Sci. USA* **93**(14): 7263–7268.

Burger, A. M., and Seth, A. K. (2004). The ubiquitin-mediated protein degradation pathway in cancer: therapeutic implications, *Eur. J. Cancer* **40**: 2217–2229.

Burnstein, K. L. (2005). Regulation of androgen receptor levels: implications for prostate cancer progression and therapy, *J. Cell Biochem.* **95**: 657–669.

Buttyan, R., Ghafar, M. A., and Shabsigh, A. (2000). The effects of androgen deprivation on the prostate gland: cell death mediated by vascular regression, *Curr. Opin. Urol.* **10**(5): 415.

Buttyan, R., Shabsigh, A., Perlman, H., *et al.* (1999). Regulation of apoptosis in the prostate gland by androgenic steroids, *Trends Endocrinol. Metab.* **10**(2): 47.

Buttyan, R., Zakeri, Z., Lockshin, R., *et al.* (1988). Cascase induction of c-fos, c-myc, and heat shock 70K transcripts during regression of the rat ventral prostate gland, *Molecular Endocrinology* **2**(7): 650–657.

Byrne, R. L., Leung, H., and Neal, D. E. (1996). Peptide growth factors in the prostate as mediators of stromal epithelial interaction, *Br. J. Urol.* **77**(5): 627–633.

Califice, S., Castronovo, V., Bracke, M., *et al.* (2004). Dual activities of galectin-3 in human prostate cancer: tumor suppression of nuclear galectin-3 vs tumor promotion of cytoplasmic galectin-3, *Oncogene* **23**(45): 7527–7536.

Camps, J. L., Chang, S. M., Hsu, T. C., *et al.* (1990). Fibroblast-mediated acceleration of human epithelial tumor growth *in vivo, Proc. Natl. Acad. Sci. USA* **87**(1): 75–79.

Cao, Y. J., Caffo, B., Choi, L., *et al.* (2008). Noninvasive quantification of drug concentration in prostate and seminal vesicles: improvement and validation with desipramine and aspirin, *J. Clin. Pharmacol.* **48**: 176–183.

Cardillo, M. R., Petrangeli, E., Perracchio, L., *et al.* (2000). Transforming growth factor-beta expression in prostate neoplasia, *Anal. Quant. Cytol. Histol.* **22**(1): 1–10.

Carducci, M. A., Saad, F., Abrahamsson, P., *et al.* (2007). A phase 3 randomized controlled trial of the efficacy and safety of atrasentan in men with metastatic hormone-refractory prostate cancer, *Cancer* **110**: 1959–1966.

Carew, J. S., Medina, E. C., Esquivel, J. A. 2nd *et al.* (2010). Autophagy inhibition enhances vorinostat-induced apoptosis via ubiquitinated protein accumulation, *J. Cell Mol. Med.* **14**(10): 2448–2459.

Carles, J., Font, A., Mellado, B., *et al.* (2007). Weekly administration of docetaxel in combination with estramustine and celecoxib in patients with advanced hormone-refractory prostate cancer: final results from a phase II study, *Br. J. Cancer* **97**: 1206–1210.

Carmeliet, P. (2000). Mechanisms of angiogenesis and arteriogenesis, *Nat. Med.* **6**(4): 389–395.

Cartee, L., Maggio, S. C., Smith, R., *et al.* (2003). Protein kinase C-dependent activation of the tumor necrosis factor receptor-mediated extrinsic cell death pathway underlies enhanced apoptosis in human myeloid leukemia cells exposed to Bryostatin 1 and Flavopiridol, *Mol. Cancer Ther.* **2**: 83–87.

Carter, B. S., Beatty, T. H., Steinberg, G. D., *et al.* (1992). Mendelian inheritance of familial prostate cancer, *Proc. Natl. Acad. Sci. USA* **89**(8): 3367–3371.

Carver, B. S., and Pandolfi, P. P. (2006). Mouse modeling in oncologic preclinical and translational research, *Clin. Cancer Res.* **12**: 5305–5311.

Catalona, W. J., Partin, A. W., Slawin, K. M., *et al.* (1998). Use of the percentage of free prostate-specific antigen to enhance differentiation of prostate cancer from benign prostatic disease, *JAMA* **279**(19): 1542–1547.

Cavallaro, U., and Christofori, G. (2004). Cell adhesion and signalling by cadherins and Ig-CAMs in cancer, *Nat. Rev. Cancer* **4**(2): 118–132.

Centenera, M. M., Harris, J. M., Tilley, W. D., *et al.* (2008). The contribution of different androgen receptor domains to receptor dimerization and signaling, *Mol. Endocrinol.* **22**(11): 2373–2382.

Chakraborty, M., Qiu, S. G., Vasudevan, K. M., *et al.* (2001). Par-4 drives trafficking and activation of FAS and FAS-L to induce prostate cancer cell apoptosis and tumor regression, *Cancer Res.* **61**: 7255–7263.

Chambers, A. F., and Matrisian, L. M. (1997). Changing views of the role of matrix metalloproteinases in metastasis, *J. Natl. Cancer. Inst.* **89**(17): 1260–70.

Chan, J. M., Stampfer, M. J., Giovannucci, E., *et al.* (1998a). Plasma insulin-like growth factor-I and prostate cancer risk: a prospective study, *Science* **279**: 563–566.

Chan, T. A., Morin, P., Vogelstein, B., *et al.* (1998b). Mechanisms underlying nonsteroidal antiinflammatory drug-mediated apoptosis, *Proc. Natl. Acad. Sci. USA* **95**: 681–686.

Che, M., and Grignon, D. (2002). Pathology of prostate cancer, *Cancer Metast. Rev.* 21(3–4): 381–395.

Chen, F. (2004). Endogenous inhibitors of nuclear factor-κB, an opportunity for cancer control, *Cancer Res.* 64: 8135–8138.

Chen, C. S. (2008). Mechanotransduction – a field pulling together? *J. Cell Sci.* 121: 3285–3292.

Chen, C. H., Wang, W. J., Kuo, J. C., *et al.* (2005). Bidirectional signals transduced by DAPK–ERK interaction promote the apoptotic effect of DAPK, *EMBO J.* 24: 294–304.

Chen, C. D., Welsbie, D. S., Tran, C., *et al.* (2004). Molecular determinants of resistance to antiandrogen therapy, *Nat. Med.* 10: 33–39.

Cheng, E.H., Nicholas, J., Bellows, D.S., *et al.* (1997). A BCL-2 homolog encoded by Kaposi sarcoma-associated virus, human herpesvirus 8, inhibits apoptosis but does not heterodimerize with BAX or BAK, *Proc. Natl. Acad. Sci. USA* 94(2): 690–694.

Chipuk, J. E., Cornelius, S. C., Pultz, N. J., *et al.* (2002). The androgen receptor represses transforming growth factor-beta signaling through interaction with Smad3, *J. Biol. Chem.* 277(2): 1240–1248.

Chon, J. K., Borkowski, A., Partin, A. W., *et al.* (1999). Alpha 1-adrenoceptor antagonists terazosin and doxazosin induce prostate apoptosis without affecting cell proliferation in patients with benign prostatic hyperplasia, *J. Urol.* 161: 2002–2008.

Chopra, D. P., Menard, R. E., Januszewski, J., *et al.* (2004). TNF-alpha-mediated apoptosis in normal human prostate epithelial cells and tumor cell lines, *Cancer Lett.* 203(2): 145–154.

Christofori, G. (2003). Changing neighbours, changing behaviour: cell adhesion molecule-mediated signalling during tumour progression, *EMBO J.* 22(10): 2318–2323.

Chua, B. T., Volbracht, C., Tan, K. O., *et al.* (2003). Mitochondrial translocation of cofilin is an early step in apoptosis induction, *Nat. Cell Biol.* 5: 1083–1089.

Chung, L. W., Baseman, A., Assikis, V., *et al.* (2005). Molecular insights into prostate cancer progression: the missing link of tumor microenvironment, *J. Urol.* 173: 10–20.

Chung, L. W., Chang, S. M., Bell, C., *et al.* (1989). Co-inoculation of tumorigenic rat prostate mesenchymal cells with non-tumorigenic epithelial cells results in the development of carcinosarcoma in syngeneic and athymic animals, *Int. J. Cancer* 43(6): 1179–1187.

Chung, L. W., Gleave, M. E., Hsieh, J. T., *et al.* (1991). Reciprocal mesenchymal-epithelial interaction affecting prostate tumour growth and hormonal responsiveness, *Cancer Surv.* 11: 91–121.

Chung, T. D., Mauceri, H. J., Hallahan, D. E., *et al.* (1998). Tumor necrosis factor-alpha-based gene therapy enhances radiation cytotoxicity in human prostate cancer, *Cancer Gene Ther.* 5(6): 344–349.

Clark, E. A., Golub, T. R., Lander, E. S., *et al.* (1999). Genomic analysis of metastasis reveals an essential role for RhoC, *Nature* 406: 532–535.

Clark, L. C., Combs, G. F. Jr, Turnbull, B. W., *et al.* (1996). Effects of selenium supplementation for cancer prevention in patients with carcinoma of the skin: a randomized controlled trial, *J. Am. Med. Assoc.* 276: 1957–1963.

Cleary, M. L., and Sklar, J. (1985). Nucleotide sequence of a t(14;18) chromosomal breakpoint in follicular lymphoma and demonstration of a breakpoint-cluster region near a transcriptionally active locus on chromosome 18, *Proc. Natl. Acad. Sci.* **82**(21): 7439–7443.

ClinicalTrials.gov (2008). "A Study of an Experimental Chemotherapy Combination to Treat Hormone Refractory Prostate Cancer." [Accessed Nov 2, 2010] http://clinicaltrials.gov/ct2/results?term=NCT00642018

Coffey, R. J. Jr, Shipley, G. D., and Moses, H. L. (1986). Production of transforming growth factors by human colon cancer lines, *Cancer Res.* **46**(3): 1164–1169.

Coffey, R. N. T., Watson, W. G., O'Neill, A. J., et al. (2002). Androgen-mediated resistance to apoptosis, *Prostate* **53**: 300–309.

Cohen, A. W., Hnasko, R., Schubert, W., et al. (2004). Role of caveolae and caveolins in health and disease, *Physiol. Rev.* **84**(4): 1341–1379.

Cohen, J. H., Kristal, A. R., and Stanford, J. L. (2000). Fruit and vegetable intakes and prostate cancer risk, *J. Natl. Cancer Inst.* **92**: 61–68.

Cohen, Y. C., Liu, K. S., Heyden, H. L., et al. (2007). Detection bias due to the effect of finasteride on prostate volume: a modeling approach for analysis of the Prostate Cancer Prevention Trial, *J. Natl. Cancer Inst.* **99**(18): 1366–1374.

Colombel, M., Gil Diez, S., Radvanyi, F., et al. (1996). Apoptosis in prostate cancer. Molecular basis to study hormone refractory mechanisms, *Ann. N Y Acad. Sci.* **784**: 63–69.

Colombel, M., Symmans, F., Gil, S., et al. (1993). Detection of the apoptosis-suppressing oncoprotein bcl-2 in hormone-refractory human prostate cancers, *Am. J. Pathol.* **143**: 390–400.

Condeelis, J., and Pollard, J. W. (2006). Macrophages: obligate partners for tumor cell migration, invasion, and metastasis, *Cell* **124**(2): 263–266.

Cook, N. R., Lee, I. M., Gaziano, J. M., et al. (2005). Low-dose aspirin in the primary prevention of cancer: the Women's Health Study: a randomized controlled trial, *JAMA* **294**: 47–55.

Cornford, P., Evans, J., Dodson, A., et al. (1999). Protein kinase C isoenzyme patterns characteristically modulated in early prostate cancer, *Am. J. Pathol.* **154**(1): 137–144.

Costa-Pereira, A. P., and Cotter, T. G. (1999). Camptothecin sensitizes androgen-independent prostate cancer cells to anti-FAS-induced apoptosis, *Br. J. Cancer* **80**(3/4): 371–378.

Coussens, L. M., and Werb, Z. (2002). Inflammation and cancer. *Nature* **420**(6917): 860–867.

Cox, J. S., Shamu, C. E., and Walter, P. (1993). Transcriptional induction of genes encoding endoplasmic reticulum resident proteins requires a transmembrane protein kinase, *Cell* **73**(6): 1197–1206.

Craft, N., Shostak, Y., Carey, M., et al. (1999). A mechanism for hormone-independent prostate cancer through modulation of androgen receptor signaling by the HER-2/neu tyrosine kinase, *Nat. Med.* **5**(3): 280–285.

Crawford, E. E. (2003). Epidemiology of prostate cancer. *Urology*, **62**: 3–12.

Critz, F. A., Levinson, K., Williams, W. H., et al. (1997). Prostate-specific antigen nadir of 0.5 ng/ml or less defines disease freedom for surgically staged men irradiated for prostate cancer, *Urology* **49**(5): 668–672.

Cronauer, M. V., Schulz, W. A., Ackermann, R., *et al.* (2005). Effects of WNT/β-catenin pathway activation on signaling through T-cell factor and androgen receptor in prostate cancer cell lines, *Int. J. Oncol.* **26**(4): 1033–1040.

Crowe, D. L., and Ohannessian. A. (2004). Recruitment of focal adhesion kinase and paxillin to β1 integrin promotes cancer cell migration via mitogen activated protein kinase activation, *BMC Cancer* **4**: 18.

Culig, Z., and Bartch, G. (2006). Androgen axis in prostate cancer, *J. Cell. Biochem.* **99**: 373–381.

Culig, Z., Hobisch, A., Cronauer, M. V., *et al.* (1994). Androgen receptor activation in prostatic tumor cell lines by insulin-like growth factor-I, keratinocyte growth factor, and epidermal growth factor, *Cancer Res.* **54**(20): 5474–5478.

Cunha, G. R., and Donjacour, A. A. (1989). Mesenchymal-epithelial interactions in the growth and development of the prostate, *Cancer Treat. Res.* **46**: 159–175.

Cunha, G. R., Fuji, H., Neubauer, B. L., *et al.* (1983). Epithelial-mesenchymal interactions in prostatic development. I. Morphological observations of prostatic induction by urogenital sinus mesenchyme in epithelium of the adult rodent urinary bladder, *J. Cell Biol.* **96**: 1662–1670.

Cunha, G. R., Hayward, S. W., Dahiya, R., *et al.* (1996). Smooth muscle-epithelial interactions in normal and neoplastic prostatic development, *Acta Anat. (Basel)* **155**(1): 63–72.

Cuzick, J., Otto, F., Baron, J. A., *et al.* (2009). Aspirin and non-steroidal anti-inflammatory drugs for cancer prevention: an international consensus statement, *Lancet Oncol.* **10**: 501–507.

Cvetkovic, D., Movsas, B., Dicker, A. P., *et al.* (2001). Increased hypoxia correlates with increased expression of the angiogenesis marker vascular endothelial growth factor in human prostate cancer, *Urology* **57**(4): 821–825.

Dale, K. M., Coleman, C. I., Henyan, N. N., *et al.* (2006). Statins and cancer risk: a meta-analysis, *JAMA* **295**: 74–80.

D'Amico, A. V., Whittington, R., Malkowicz, S. B., *et al.* (1998). Biochemical outcome after radical prostatectomy, external beam radiation therapy or interstitial radiation therapy for clinically localized prostate cancer, *JAMA* **280**(11): 969–974.

Danial, N. N., and Korsmeyer, S. J. (2004). Cell death: critical control points, *Cell* **116**: 205–219.

Dasgupta, K., Di Cesar, D., Ghoshn, J., *et al.* (2006). Association between nonsteroidal and inflammatory drugs and prostate cancer occurrence. *Cancer J.* **12**: 130–135.

Davies, M. A., Koul, D., Dhesi, H., *et al.* (1999). Regulation of Akt/PKB activity, cellular growth, and apoptosis in prostate carcinoma cells by MMAC/PTEN, *Cancer Res.* **59**(11): 2551–2556.

Davis, J. W., Kolm, P., Wright, G. L., *et al.* (1999). The durability of external beam radiation therapy for prostate cancer: can it be identified? *J. Urol.* **162**: 758–761.

Debatin, K. M., and Krammer, P. H. (2004). Death receptors in chemotherapy and cancer, *Oncogene* **23**(16): 2950–2966.

Debes, J. D., and Tindall, D. J. (2004). Mechanisms of androgen-refractory prostate cancer, *NEJM* **351**(15): 1488–1490.

Debnath, J., Millis, K. R., Collins, N. L., *et al.* (2002). The role of apoptosis in creating and maintaining luminal space within normal and oncogene-expressing mammary acini, *Cell* 111(1): 29–40.

De Dosso S., and Berthold D. R. (2008). Docetaxel in the management of prostate cancer: current standard of care and future directions, *Expert Opin. Pharmacother.* 9(11): 1969–1979.

Dehm, S. M., Schmidt, L. J., Heemers, H. V., *et al.* (2008). Splicing of a novel androgen receptor exon generates a constitutively active androgen receptor that mediates prostate cancer therapy resistance, *Cancer Res.* 68: 5469–5477.

Dehm, S. M., and Tindall, D. J. (2006). Ligand-independent androgen receptor activity is activation function-2-independent and resistant to antiandrogens in androgen refractory prostate cancer cells, *J. Biol. Chem.* 281: 27882–27893.

De la Taille, A., Chen, M. W., Shabsigh, A., *et al.* (1999). FAS antigen/CD-95 upregulation and activation during castration-induced regression of the rat ventral prostate gland, *Prostate* 40(2): 89–96.

Delongchamps, N. B., Peyromaure, M., and Dinh-Xuan, A. T. (2006). Role of vascular endothelial growth factor in prostate cancer, *Urology* 68(2): 244–248.

De Marzo, A. M., Marchi, V. L., Epstein, J. I., *et al.* (1999). Proliferative inflammatory atrophy of the prostate: implications for prostatic carcinogenesis, *Am. J. Pathol.* 153: 1985–1992.

De Marzo, A. M., Nelson, W. G., Meeker, A. K., *et al.* (1998). Stem cell features of benign and malignant prostate epithelial cells, *J. Urol.* 160(6:2): 2381–2392.

De Marzo, A. M., Platz, E. A., Sutcliffe, S., *et al.* (2007). Inflammation in prostate carcinogenesis, *Nat. Rev. Cancer* 7: 256–269.

Denmeade, S. R., and Isaacs, J. T. (2002). A history of prostate cancer treatment, *Nat. Rev.* 2: 389–2396.

Denmeade, S. R., and Isaacs, J. T. (2005). The SERCA pump as therapeutic target, *Cancer Biol. Ther.* 4(1): 14–22.

Dennis, L. K., Lynch, C. F., and Tomer, J. C. (2002). Epidemiologic association between prostatitis and prostate cancer, *Urology* 60: 78–83.

Derynck, R., Akhurst, R. J., and Balmain, A. (2001). TGF-β signaling in tumor suppression in cancer progression, *Nat. Genet.* 29: 117–129.

Derynck, R., and Zhang, Y. E. (2003). Smad-dependent and Smad-independent pathways in TGF-beta family signaling, *Nature* 425(6958): 577–584.

DesMarais, V., Ghosh, M., Eddy, R., *et al.* (2005). Cofilin takes the lead, *J. Cell Sci.* 118: 19–26.

Dhanabal, M., Volk, R., Ramchandran, R., *et al.* (1999). Cloning, expression, and *in vitro* activity of human endostatin, *Biochem. Biophys. Res. Commun.* 258(2): 345–352.

Dhanalakshmi, S., Singh, R. P., Agarwal, C., *et al.* (2002). Silibinin inhibits constitutive and TNFalpha-induced activation of NF-kappaB and sensitizes human prostate carcinoma DU145 cells to TNFalpha-induced apoptosis, *Oncogene* 21(11): 1759–1767.

Diaz-Montero, C. M., Wygant, J. N., and McIntyre, B. W. (2006). PI3-K/Akt-mediated anoikis resistance of human osteosarcoma cells requires Src activation, *Eur. J. Cancer* 42: 1491–1500.

DiGiovanni, J., Kiguchi, K., Frijhoff, A., *et al.* (2000). Deregulated expression of insulin-like growth factor 1 in prostate epithelium leads to neoplasia in transgenic mice, *Proc. Natl. Acad. Sci. USA* **97**(7): 3455–3460.

Di Lorenzo, G., Tortora, G., D'Armiento, F. P., *et al.* (2002). Expression of epidermal growth factor receptor correlates with disease relapse and progression to androgen-independence in human prostate cancer, *Clin. Cancer Res.* **8**(11): 3438–3444.

Ding, Z., Wu, C-J., Chu, G., *et al.* (2011). SMAD4-dependent barrier constrains prostate cancer growth and metastatic progression, *Nature* **470**: 269–273.

DiNitto, J. P., Cronin, T. C., and Lambright, D. G. (2003). Membrane recognition and targeting by lipid-binding domains, *Sci. STKE* **213**: re16.

Dionne, C. A., Camoratto, A. M., Jani, J. P., *et al.* (1998). Cell cycle-independent death of prostate adenocarcinoma is induced by the trk tyrosine kinase inhibitor CEP-751 (KT6587), *Clin. Cancer Res.* **4**(8): 1887–1898.

DiPaola, R. S., Patel, J., and Rafi, M. M. (2001). Targeting apoptosis in prostate cancer, *Hematol. Oncol. Clin. North Am.* **15**: 509–524.

Diwadkar-Navsariwala, V., Prins, G. S., Swanson, S. M., *et al.* (2006). Selenoprotein deficiency accelerates prostate carcinogenesis in a transgenic model, *Proc. Natl. Acad. Sci.* **103**(21): 8179–8184.

Dixelius, J., Cross, M., Matsumoto, T., *et al.* (2002). Endostatin regulates endothelial cell adhesion and cytoskeletal organization, *Cancer Res.* **62**(7): 1944–1947.

Djakiew, D. (2000). Deregulated expression of growth factors and their receptors in the development of prostate cancer, *Prostate* **38**: 268–277.

Djakiew, D., Lamb, J., Bova, S., *et al.* (1998). Cell cycle-independent death of prostate adenocarcinoma is induced by the trk tyrosine kinase inhibitor CEP-751 (KT6587), *Clin. Cancer Res.* **4**: 1887–1898.

Djavan, B., Waldert, M., Seitz, C., *et al.* (2001). Insulin-like growth factors and prostate cancer, *World J. Urol.* **19**(4): 225–233.

Doll, J. A., Reiher, F. K., Crawford, S. E., *et al.* (2001). Thrombospondin-1, vascular endothelial growth factor and fibroblast growth factor-2 are key functional regulators of angiogenesis in the prostate, *Prostate* **49**(4): 293–305.

Dong, B., Kim, S., Hong, S., *et al.* (2007). An infectious retrovirus susceptible to an IFN antiviral pathway from human prostate tumors, *Proc. Natl. Acad. Sci. USA* **104**: 1655–1660.

Douma, S., Van Laar, T., Zevenhoven, J., *et al.* (2004). Suppression of anoikis and induction of metastasis by the neurotrophic receptor TrkB, *Nature* **430**(7003): 1034–1039.

Dourbin, N., Bhatt, A. K., Dutt, P., *et al.* (2001). Reduced cell migration and disruption of the actin cytoskeleton in calpain-deficient embryonic fibroblasts, *J. Biol. Chem.* **276**: 48382–48388.

Dow, J. K., and deVere White, R. W. (2000). Fibroblast growth factor 2: its structure and property, paracrine function, tumor angiogenesis, and prostate-related mitogenic and oncogenic functions, *Urology* **55**(6): 800–806.

Drake, M. J., Robson, W., Mehta, P., *et al.* (2003). An open-label phase II study of low-dose thalidomide in androgen-independent prostate cancer, *British Journal of Cancer* **88**: 822–827.

Dreicer, R., Klein, E. A., Elson, P., *et al.* (2005). Phase II trial of GM-CSF + thalidomide in patients with androgen-independent metastatic prostate cancer, *Urologic Oncology: Seminars and Original Investigations* **23**(2): 82–86.

Du, C., Fang, M., Li, Y., *et al.* (2000). SMAC, a mitochondrial protein promotes cytochrome C-dependent caspase activation by eliminating IAP inhibition, *Cell* **102**: 33–42.

Duffield-Lillico, A. J., Slate, E. H., Reid, M. E., *et al.* (2003). Selenium supplementation and secondary prevention of nonmelanoma skin cancer in a randomized trial, *J. Natl. Cancer Inst.* **95**: 1477–1481.

Duksin, D., and Mahoney, W. C. (1982). Relationship of the structure and biological activity of the natural homologues of tunicamycin, *J. Biol. Chem.* **257**(6): 3105–3109.

Dupont, J., Fernandez, A. M., Glackin, C. A., *et al.* (2001). Insulin-like growth factor 1 (IGF-1)-induced twist expression is involved in the anti-apoptotic effects of the IGF-1 receptor, *J. Biol. Chem.* **276**(28): 26699–26707.

Duxbury, M. S., Ito, H., Benoit, E., *et al.* (2004). A novel role for carcinoembryonic antigen-related cell adhesion molecule 6 as a determinant of gemcitabine chemoresistance in pancreatic adenocarcinoma cells, *Cancer Res.* **64**: 3987–3993.

Dvorak, H. F., Brown, L. F., Detmar, M., *et al.* (1995). Vascular permeability factor/vascular endothelial growth factor, microvascular hyperpermeability, and angiogenesis, *Am. J. Pathol.* **146**(5): 1029–39.

Edlund, S., Bu, S., Schuster, N., *et al.* (2003). Transforming growth factor-β1 (TGF-β)-induced apoptosis of prostate cancer cells involves Smad7-dependent activation of p38 by TGF-β- activated kinase 1 and mitogen-activated protein kinase kinase 3, *Molecular Biology of Cell* **14**(2): 529–544.

Edlund, S., Lee, S. Y., Grimsby, S., *et al.* (2005). Interaction between Smad7 and β-Catenin: Importance for transforming growth factor β-induced apoptosis, *Mol. Cell. Biol.* **25**(4): 1475–1488.

Edwards, J., Krishna, N. S., Witton, C. J., *et al.* (2003). Gene amplifications associated with the development of hormone-resistant prostate cancer, *Clin. Cancer Res.* **9**: 5271–5281.

Eger, A., Stockinger, A., Schaffhauser, B., *et al.* (2000). Epithelial mesenchymal transition by c-Fos estrogen receptor activation involves nuclear translocation of β-catenin and upregulation of β-catenin/lymphoid enhancer binding factor-1 transcriptional activity, *J. Cell Biol.* **148**(1): 173–187.

Eisenberger, M. A., Blumenstein, B. A., Crawford, E. D. *et al.*, (1998). Bilateral orchiectomy with or without flutamide for metastatic prostate cancer, *N. Engl. J. Med.* **339**: 1036–1042.

Eisenberg-Lerner, A., Bialik, S., Simon, H. U., *et al.* (2009). Life and death partners: apoptosis, autophagy and the cross-talk between them, *Cell Death Differ.* **16**: 966–975.

Ellis, H. M., and Horvitz, H. R. (1986). Genetic control of programmed cell death in the nematode *C. elegans*, *Cell* **44**(6): 817.

Enders, A., Bouillet, P., Puthalakath, H., *et al.* (2003). Loss of the pro-apoptotic BH3-only BCL-2 family member BIM inhibits BCR stimulation-induced apoptosis and deletion of autoreactive B cells, *J. Exp. Med.* **198**(7): 1119–1126.

Enenstein, J., Waleh, N.S., and Kramer, R. H. (1992). Basic FGF and TGF-β differentially modulate integrin expression of human microvascular endothelial cells, *Exp. Cell Res.* 203(2): 499–503.

English, H. F., Drago, J. R., and Santen, R. J. (1985). Cellular response to androgen depletion and repletion in the rat ventral prostate: autoradiography and morphometric analysis, *Prostate* 7(1): 41.

English, H. F., Kyprianou, N., and Isaacs, J. T. (1989). Relationship between DNA fragmentation and apoptosis in the programmed cell death in the rat prostate following castration, *Prostate* 15(3): 233–250.

English, H. F., Santen, R. J., and Isaacs, J. T. (1987). Response of glandular versus basal rat ventral prostatic epithelial cells to androgen withdrawal and replacement, *Prostate* 11(3): 229–242.

Erdreich-Epstein, A., Shimada, H., Groshen, S., et al. (2000). Integrins αvβ3 and αvβ5 are expressed by endothelium of high-risk neuroblastoma and their inhibition is associated with increased endogenous ceramide, *Cancer Res.* 60: 712.

Erler, J. T., Bennewith, K. L., Cox, T. R., et al. (2009). Hypoxia-induced lysyl oxidase is a critical mediator of bone marrow cell recruitment to form the premetastatic niche, *Cancer Cell* 15: 35–44.

Espina, V., Wulfkuhle, J. D., Calvert, V. S., et al. (2006). Laser capture micro-dissection, *Nat. Protoc.* 1(2): 586–603.

Evans, G. S., and Chandler, J. A. (1987). Cell proliferation studies in rat prostate. I. The proliferative role of basal and secretory epithelial cells during normal growth, *Prostate* 10(2): 163.

Fan, W., Yanase, T., Morinaga, H., et al. (2007). Insulin-like growth factor 1/insulin signaling activates androgen signaling through direct interactions of Foxo1 with androgen receptor, *J. Biol. Chem.* 282(10): 7329–7338.

Farhana, L., Dawson, M. I., Huang, Y., et al. (2004). Apoptosis signaling by the novel compound 3-Cl-AHPC involves increased EGFR proteolysis and accompanying decreased phosphatidylinositol 3-kinase and AKT kinase activities, *Oncogene* 23(10): 1874–1884.

Farwell, W. R., Scranton, R. E., Lawler, E. V. et al. (2008). The association between statins and cancer incidence in a veterans population, *J. Natl. Cancer Inst.* 100: 134–139.

Feldman, B. J., and Feldman, D. (2001). The development of androgen-independent prostate cancer, *Nat. Rev. Cancer* 1(1): 34–45.

Felton, J. S., Knize, M. G., Wu, R. W., et al. (2007). Mutagenic potency of food-derived heterocyclic amines, *Mutat. Res.* 616: 90–94.

Ferrara, N., Carver-Moore, K., Chen, H., et al. (1996). Heterozygous embryonic lethality induced by targeted inactivation of the VEGF gene, *Nature* 380: 439–442.

Ferrara, N., Gerber, H. P., and LeCouter, J. (2003). The biology of VEGF and its receptors, *Nat. Med.* 9: 669–676.

Ferri, K. F., and Kroemer, G. (2001). Organelle-specific initiation of cell death pathways, *Nat. Cell Biol.* 3(11): E255–263.

Festuccia, C., Muzi, P., Millimaggi, D., et al. (2005). Molecular aspects of gefitinib antiproliferative and pro-apoptotic effects in PTEN-positive and PTEN-negative prostate cancer cell lines, *Endocr. Relat. Cancer* 12(4): 983–998.

Fizazi, K. (2007). The role of Src in prostate cancer, *Ann. Oncol.* 18: 1765–1773.

Fizazi, K., Yang, J., Peleg, S., *et al.* (2003). Prostate cancer cells-osteoblast interaction shifts expression of growth/survival-related genes in prostate cancer and reduces expression of osteoprotegerin in osteoblasts, *Clin. Cancer Res.* 9: 2587–2597.

Fleisch, M. C., Maxwell, C. A., and Barcellos-Hoff, M-H. (2006). The pleiotropic roles of transforming growth factor beta in homeostasis and carcinogenesis of endocrine organs, *Endocr. Relat. Cancer* 13(2): 379–400.

Fleisher, M., Danila, D. C., Leversha, M., *et al.* (2009). Circulating tumor cells (CTC) in patients with metastatic castration-resistant prostate cancer (CRPC) receiving abiraterone acetate (AA) after failure of docetaxel-based chemotherapy, *J. Clin. Oncol.* 27(15s): 5049.

Flick, E. D., Habel, L. A., Chan, K. A., *et al.* (2007). Statin use and risk of prostate cancer in the California men's health study cohort, *Cancer Epidem. Biom.* 16: 2218.

Folkman, J. (1971). Tumor angiogenesis: therapeutic implications, *N. Eng. J. Med.* 285: 1182–1186.

Folkman, J. (1993). 'Tumor angiogenesis', in Holland, J. F., Frei, E. III, Bast, R. C. Jr, Kufe, D. W., Morton, D. L., and Weichselbaum, R. R. (eds), *Cancer Medicine*, Lea & Febiger, Philadelphia, vol. 1, 3rd ed., pp. 153–170.

Folkman, J. (1995a). Angiogenesis in cancer, vascular, rheumatoid and other disease, *Nat. Med.* 1: 27–31.

Folkman, J. (1995b). Clinical applications of research on angiogenesis, *NEJM.* 333(26): 1757–1763.

Fong, G. H., Rossant, J., Gertsenstein, M., *et al.* (1995). Role of the flt-1 receptor tyrosine kinase in regulating the assembly of vascular endothelium, *Nature* 376(6535): 66–70.

Fong, T. A., Shawver, L. K., Sun, L., *et al.* (1999). SU5416 is a potent and selective inhibitor of the vascular endothelial growth factor receptor (Flk-1/KDR) that inhibits tyrosine kinase catalysis, tumor vascularization, and growth of multiple tumor types, *Cancer Res.* 59: 99–106.

Fornaro, M., Manes, T., and Languino, L. R. (2001). Integrin and prostate cancer metastases, *Cancer Metast. Rev.* 20: 321–331.

Franck-Lissbrant, I., Häggström, S., Damber, J. E., *et al.* (1998). Testosterone stimulates angiogenesis and vascular regrowth in the ventral prostate in castrated adult rats, *Endocrinology* 139(2): 451–456.

Freeman, M. R., Cinar, B., Kim, J., *et al.* (2007). Transit of hormonal and EGF receptor-dependent signals through cholesterol-rich membranes, *Steroids* 72(2): 210–217.

Frisch, S. M., and Francis, H. (1994). Disruption of epithelial cell-matrix interactions induces apoptosis, *J. Cell Biol.* 124(4): 619–626.

Frisch, S. M., and Ruoslahti, E. (1997). Integrins and anoikis, *Curr. Opin. Cell Biol.* 9: 701–706.

Frisch, S. M., and Screaton, R. A. (2001). Anoikis mechanisms, *Curr. Opin. Cell Biol.* 13(5): 555–562.

Frisch, S. M., Vuori, K., Ruoslahti, E., *et al.* (1996). Control of adhesion-dependent cell survival by focal adhesion kinase. *J. Cell Biol.* 134: 793–799.

Frost, P., Pei Ng, C., Belldegrun, A., and Bonavida, B. (1997). Immunosensitization of prostate carcinoma cell lines for lymphocytes (CTL, TIL, LAK)-mediated apoptosis via the FAS-FAS-ligand pathway of cytotoxicity, *Cell. Immunol.* 180(1): 70–83.

Fukuda, T., Chen, K., Shi, X., *et al.* (2003). PINCH-1 is an obligate partner of integrin-linked kinase (ILK) functioning in cell shape modulation, motility, and survival, *J. Biol. Chem.* **278**: 51324–51333.

Fung, C., Lock, R., Gao, S., *et al.* (2008). Induction of autophagy during extracellular matrix detachment promotes cell survival, *Mol. Biol. Cell*, **19**: 797–806.

Furuya, Y., Krajewski, S., Epstein, J. I., *et al.* (1996). Expression of bcl-2 and the progression of human and rodent prostatic cancers, *Clin. Cancer Res.* **2**: 389–398.

Gabbiani, G., and Majno, G. (1972). Dupuytren's Contracture: Fibroblast Contraction? An Ultrastructural Study, *Am. J. Pathol.* **66(1)**: 131–146.

Galaup, A., Opolon, P., Bouquet, C., *et al.* (2003). Combined effects of docetaxel and angiostatin gene therapy in prostate tumor model, *Mol. Ther.* **7(6)**: 731–740.

Gamble, S. C., Odontiadis, M., Waxman, J., *et al.* (2004). Androgen target prohibitin to regulate proliferation of prostate cancer cells, *Oncogene* **23**: 2996–3004.

Gan, L., Chen, S., Wang, Y., *et al.* (2009). Inhibition of the androgen receptor as a novel mechanism of taxol chemotherapy in prostate cancer, *Cancer Res.* **69**: 8386–8394.

Gao, N., Zhang, Z., Jiang, B. H., *et al.* (2003). Role of PI3K/AKT/mTOR signaling in the cell cycle progression of human prostate cancer, *Biochem. Biophys. Res. Commun.* **310(4)**: 1124–1132.

Garcia-Moreno, C., Mendez-Davila, C., de La Piedra, C., *et al.* (2002). Human prostatic carcinoma cells produce an increase in the synthesis of interleukin-6 by human osteoblasts, *Prostate* **50(4)**: 241–246.

Garcia Rodriguez, L. A., and González-Pérez, A. (2004). Inverse association between nonsteroidal anti-inflammatory drugs and prostate cancer, *Cancer Epidem. Biom.* **13**: 649–653.

Garrison, J. B., and Kyprianou, N. (2004). Novel targeting of apoptosis pathways for prostate cancer therapy, *Curr. Cancer Drug Targets* **4(1)**: 85–95.

Garrison, J. B., and Kyprianou, N. (2006). Doxazosin induces apoptosis of benign and malignant prostate cells via a death receptor-mediated pathway, *Cancer Res.* **66**: 464–472.

Garrison, J. B., Shaw, Y. J., Chen, C. S., *et al.* (2007). Novel quinazoline-based compounds impair prostate tumorigenesis by targeting tumor vascularity, *Cancer Res.* **67**: 11344–11352.

Gaziano, J. M., Glynn, R. J., Christen, W. G., *et al.* (2009). Vitamins E and C in the prevention of prostate and total cancer in men: the Physician's Health Study II randomized controlled trial. *JAMA*, **301**: 52–62.

Geller, J. (1990). Effect of finasteride, a 5α-reductase inhibitor on prostate tissue androgens and prostate-specific antigen, *J. Clin. Endocrinol. Metabol.* **71**: 1552–1555.

George, D. J., Dionne, C. A., Jani, J., *et al.* (1999). Sustained *in vivo* regression of Dunning H rat prostate cancers treated with combinations of androgen ablation and Trk tyrosine kinase inhibitors, CEP-751 (KT-6587) or CEP-701 (KT-5555), *Cancer Res.* **59(10)**: 2395–2401.

George, D. J., Halabi, S., Shepard, T. F., *et al.* (2005). The prognostic significance of plasma interleukin-6 levels in patients with metastatic hormone-refractory prostate cancer: results from cancer and leukemia group B 9480, *Clin. Cancer Res.* **11**: 1815–1820.

Gerdes, M. J., Dang, T. D., Larsen, M., *et al.* (1998). Transforming growth factor-beta1 induces nuclear to cytoplasmic distribution of androgen receptor and inhibits androgen response in prostate smooth muscle cells, *Endocrinology* **139**(8): 3569–3577.

Gerdes, M. J., Larsen, M., Dang, T. D., *et al.* (2004). Regulation of rat prostate stromal cell myodifferentiation by androgen and TGF-beta1, *Prostate* **58**(3): 299–307.

Ghafar, M. A., Anastasiadis, A. G., Chen, M. W., *et al.* (2003). Acute hypoxia increases the aggressive characteristics and survival properties of prostate cancer cells, *Prostate* **54**(1): 58–67.

Ghosh, P. M., Malik, S., Bedolla, R., *et al.* (2003). Akt in prostate cancer: possible role in androgen-independence, *Curr. Drug Metab.* **4**(6): 487–496.

Giancotti, F. G., and Ruoslahti, E. (1999). Integrin signaling, *Science* **285**: 1028–1032.

Giannakakou, P., Sackett, D., and Fojo, T. (2000). Tubulin/microtubules: still a promising target for new chemotherapeutic agents. *J. Natl. Cancer Inst.* **92**: 182–183.

Giannakakou, P., Sackett, D. L., Ward, Y., *et al.* (2000). p53 is associated with cellular microtubules and is transported to the nucleus by dynein, *Nat. Cell Biol.* **2**: 709–717.

Giannoni, E., Buricchi, F., Grimaldi, G., *et al.* (2008). Redox regulation of anoikis: reactive oxygen species as essential mediators of cell survival, *Cell Death Differ.* **15**: 867–878.

Giannoni, E., Fiaschi, T., Ramponi, G., *et al.* (2009). Redox regulation of anoikis resistance of metastatic prostate cancer cells: key role for Src and EGFR-mediated pro-survival signals oxidants confer anoikis resistance to cancer cells, *Oncogene* **28**: 2074–2086.

Gilley, J., Coffer, P. J., and Ham, J. (2003). FOXO transcription factors directly activate bim gene expression and promote apoptosis in sympathetic neurons, *J. Biol. Chem.* **162**(4): 613–622.

Gingrich, J. R., Barrios, R. J., Kattan, M. W., *et al.* (1997). Androgen-independent prostate cancer progression in the TRAMP model, *Cancer Res.* **57**: 4687–4691.

Glass, T. R., Tangen, C. M., Crawford, E. D., *et al.* (2003). Metastatic carcinoma of the prostate: identifying prognostic groups using recursive partitioning, *J. Urol.* **169**(1): 164–169.

Gleason, D. F. (1966). Classification of prostatic carcinomas, *Cancer Chemoth. Rep.* **50**: 125–128.

Gleason, D. F., and Mellinger, G. T. (1974). Prediction of prognosis for prostatic adenocarcinoma by combined histological grading and clinical staging, *J. Urol.* **111**: 58–84.

Gleave, M., Hsieh, J. T., Gao, C., *et al.* (1991). Acceleration of human prostate cancer growth *in vivo* by factors produced by prostate and bone fibroblasts, *Cancer Res.* **51**(14): 3753–3761.

Gleave, M., Tolcher, A., Miyake, H., *et al.* (1999). Progression to androgen independence is delayed by adjuvant treatment with antisense BCL-2 oligodeoxynucleotides after castration in the LNCaP prostate tumor model, *Clin. Cancer Res.* **5**: 2891–2898.

Glickman. M. H., and Ciechanover, A. (2002). The ubiquitin-proteasome proteolytic pathway: Destruction for the sake of construction, *Physiological Reviews* **82**(2): 373–428.

Gloyan, R. E., Siiteri, P. K., and Wilson, J. D. (1970). Dihydrotestosterone in prostatic hypertrophy. II. The formation and content of dihydrotestosterone in the hypertrophic canine prostate and the effect of dihydrotestosterone on prostate growth in the dog, *J. Clin. Invest.* **49(9)**: 1746.

Goel, H. L., Breen, M., Zhang, J., *et al.* (2005). β1A integrin expression is required for type 1 insulin-like growth factor receptor mitogenic and transforming activities and localization to focal contacts, *Cancer Res.* **65**: 6692.

Goel, H. L., and Languino, L. R. (2004). Integrin signaling in cancer, *Cancer Treat. Res.* **119**: 15–31.

Goel, H. L., Li, J., Kogan, S., and Languino, L. R. (2008). Integrins in prostate cancer progression, *Endocr. Relat. Cancer* **15**: 657–664.

Gold, L. (1999). The role of transforming growth factor-beta (TGF-beta) in human cancer, *Crit. Rev. Oncology.* **10(4)**: 303–360.

Gonzalez-Herrera, I. G., Prado-Lourenco, L., Pileur, F., *et al.* (2006). Testosterone regulates FGF-2 expression during testis maturation by an IRES-dependent translational mechanism, *FASEB J.* **20(3)**: 476–478.

Goodman, P. J., Thompson, I. M., Tangen, C. M., *et al.* (2006). The prostate cancer prevention trial: design, biases and interpretation of study results, *J. Urol.* **175(6)**: 2234–2242.

Goswani, A., Burikhanov, R., de Thonel, A., *et al.* (2005). Binding and phosphorylation of par-4 by akt is essential for cancer cell survival, *Mol. Cell* **20**: 33–44.

Graff, J. R., Deddens, J. A., Konicek, B. W., *et al.* (2001). Integrin-linked kinase expression increases with prostate tumor grade, *Clin. Cancer Res.* **7**: 1987–1991.

Grassme, H., Cremesti, A., Kolesnick, R., *et al.* (2003). Ceramide-mediated clustering is required for CD95-DISC formation, *Oncogene* **22(35)**: 5457–5470.

Gray, D. C., Mahrus, S., and Wells, J. A. (2010). Activation of specific apoptotic caspases with an engineered small-molecule-activated protease, *Cell* **142**: 637–646.

Gregory, C. W., Fei, X., Ponguta, L. A., *et al.* (2004). Epidermal growth factor increases coactivation of the androgen receptor in recurrent prostate cancer, *J. Biol. Chem.* **279(8)**: 7119–7130.

Gregory, C. W., Hamil, K. G., Kim, D., *et al.* (1998). Androgen receptor expression in androgen-independent prostate cancer is associated with increased expression of androgen-regulated genes, *Cancer Res.* **58**: 5718–5724.

Gregory, C. W., He, B., Johnson, R. T., *et al.* (2001a). A mechanism for androgen receptor-mediated prostate cancer recurrence after androgen deprivation therapy, *Cancer Res.* **61**: 4315.

Gregory, C. W., Johnson, R. T., Mohler, J. L., *et al.* (2001b). Androgen receptor stabilization in recurrent prostate cancer is associated with hypersensitivity to low androgen, *Cancer Res.* **61**: 2892.

Greenberg, N. M., DeMayo, F., Finegold, M. J., *et al.* (1995). Prostate cancer in a transgenic mouse, *Proc. Natl. Acad. Sci. USA* **92**: 3439–3443.

Greenblatt, M., and Shubick, P. (1968). Tumor angiogenesis: transfilter diffusion studies in the hamster by the transparent chamber technique, *J. Natl. Cancer Inst.* **41**: 111–124.

Greten, F. R., Eckman, L., Greten, T. F., *et al.* (2004). IKKB links inflammation and tumorigenesis in a mouse model of colitis-associated cancer, *Cell* **118**: 285–296.

Grignon, D. J., Caplan, R., Sarker, F. H., *et al.* (1997). p53 status and prognosis of locally advanced prostatic adenocarcinoma: a study based on RTOG 8610, *J. Natl. Cancer Inst.* **89**(2): 158–165.

Grimm, D., Streetz, K. L., Jopling, C. L., *et al.* (2005). Fatality in mice due to oversaturation of cellular microRNA/short hairpin RNA pathways, *Nature* **441**: 537–541.

Grossfeld, G. D., Small. E. J., and Carroll, P. R. (1998). Intermittent androgen deprivation for clinically localized prostate cancer: initial experience, *Urology* **51**(1): 137–144.

Grossmann, M. E., Huang, H., and Tindall, D. J. (2001). Androgen receptor signaling in androgen-refractory prostate cancer, *J. Natl. Cancer Inst.* **93**(22): 1687–1697.

Gudkov, A. V. (2003). Cooperation of two mutant p53 alleles contributes to FAS resistance of prostate carcinoma cells, *Cancer Res.* **63**(11): 2905–2912.

Gullberg, D., Gehlsen, K. R., Turner, D. C., *et al.* (1992). Analysis of alpha 1 beta 1, alpha 2 beta 1 and alpha 3 beta 1 integrins in cell-collagen interactions: identification of conformation dependent alpha 1 beta 1 binding sites in collagen type I, *EMBO J.* **11**: 3865–3873.

Gunawardena, K., Murray, D. K., and Meikle, A. W. (2000). Vitamin E and other antioxidants inhibit human prostate cancer cells through apoptosis, *Prostate* **44**: 287–295.

Guo, Y., and Kyprianou, N. (1998). Overexpression of transforming growth factor (TGF)-β1 type II receptor, restores TGF-β1-sensitivity and signaling in human prostate cancer cells, LNCaP, *Cell Growth Differ.* **9**: 185–193.

Guo, Y., and Kyprianou, N. (1999). Restoration of transforming growth factor beta signaling pathway in human prostate cancer cells suppresses tumorigenicity via induction of caspase-1-mediated apoptosis, *Cancer Res.* **59**(6): 1366–1371.

Guo, Z., Yang, X., Sun, F., *et al.* (2009). A novel androgen receptor splice variant is up-regulated during prostate cancer progression and promotes androgen depletion-resistant growth, *Cancer Res.* **69**: 2305–2313.

Gupta, G. P., and Massague, J. (2006). Cancer metastasis: building a framework, *Cell* **127**: 679–695.

Gupta, S., Hastak, K, Ahmad, N., *et al.* (2001). Inhibition of prostate carcinogenesis in TRAMP mice by oral infusion of green tea polyphenols, *Proc. Natl. Acad. Sci. USA* **98**: 10350–10355.

Gurova, K. V., Rokhlin, O. W., Budanov, A. V., *et al.* (2005). Phosphorylation of Par-4 by protein kinase A is critical for apoptosis, *Mol. Cell Biol.* **25**: 1146–1161.

Gurumurthy, S., Goswami, A., Vasudevan, K. M., and Rangnekar, V. M. (2005). Phosphorylation of par-4 by protein kinase A is critical for apoptosis, *Mol. Cell. Biol.* **25**(3): 1146–1161.

Guseva, N. V., Taghiyev, A. F., Rokhlin, O. W., *et al.* (2004). Death receptor-induced cell death in prostate cancer, *J. Cell Biochem.* **91**(1): 70–99.

Gustin, J. A., Maehama, T., Dixon, J. E., *et al.* (2001). The PTEN tumor suppressor protein inhibits tumor necrosis factor-induced nuclear factor kappa B activity, *J. Biol. Chem.* **276**(29): 27740–27744.

Haggstorm, S., Wikstrom, P., Bergh, A., *et al.* (1998). Expression of vascular endothelial growth factor and its receptors in the rat ventral prostate and Dunning R3327 PAP adenocarcinoma before and after castration, *Prostate* **36**(2): 71–75.

Haldar, S., Chintapalli, J., and Groce, C. M. (1996). Taxol induces bcl-2 phosphorylation and death of prostate cancer cells, *Cancer Res.* **56**: 1253–1255.

Haldar, S., Jena, N., and Croce, C. M. (1995). Inactivation of bcl-2 by phosphorylation, *Proc. Nati. Acad. Sci. USA* **92**: 4507–4511.

Halkidou, K., Gaughan, L., Cook, S., *et al.* (2004). Upregulation and nuclear recruitment of HDAC1 in hormone refractory prostate cancer, *Prostate* **59**: 177–189.

Hall, C. L., Dai, J., van Golen, K. L., *et al.* (2006). Type I collagen receptor (α2β1) signaling promotes the growth of human prostate cancer cells within the bone, *Cancer Res.* **66**: 8648–8654.

Hamilton, R. J., Goldberg, K. C., Platz, E. A., *et al.* (2008). The influence of statin medications on prostate-specific antigen levels, *J. Natl. Cancer Inst.* **100**: 11511–11518.

Hammes, S. R., and Levin E. R. (2007). Extranuclear steroid receptors: nature and function, *Endocr. Rev.* **28**: 726–741.

Hammond, J. W., Cai, D., and Verhey, K. J. (2008). Tubulin modifications and their cellular functions, *Curr. Opin. Cell Biol.* **20**: 71–76.

Hampton, R. Y. (2002). ER-associated degradation in protein quality control and cellular regulation, *Curr. Opin. Cell Biol.* **14**(4): 476–482.

Hanada, M., Aime-Sempe, C., Sato, T., and Reed, J. C. (1995). Structure-function analysis of Bcl-2 protein. Identification of conserved domains important for homo-dimerization with BCL-2 and heterodimerization with BAX, *J. Biol. Chem.* **270**: 11962–11969.

Hanahan, D., and Folkman, J. (1996). Patterns and emerging mechanisms of the angio-genic switch during tumorigenesis, *Cell* **86**(3): 353–364.

Hanahan, D., and Weinberg, R. A. (2000). The hallmarks of cancer, *Cell* **100**: 57–70.

Handsley, M. M., and Edwards, D. R. (2005). Metalloproteinases and their inhibitors in tumor angiogenesis, *Int. J. Cancer* **115**(6): 849–860.

Hannigan, G., Troussard, A. A., and Dedhar, S. (2005). Integrin-linked kinase: a cancer therapeutic target unique among its ILK, *Nat. Rev. Cancer* **5**: 51–63

Hannigan, G. E., Leung-Hagesteijn, C., Fitz-Gibbon, L., *et al.* (1996). Regulation of cell adhesion and anchorage-dependent growth by a new beta 1-integrin-linked protein kinase, *Nature* **379**: 91–96.

Harada, S., Keller, E. T., Fujimoto, N., *et al.* (2001). Long-term exposure of tumor necrosis factor alpha causes hypersensitivity to androgen and anti-androgen withdrawal phenomenon in LNCaP prostate cancer cells, *Prostate* **46**(4): 319–326.

Harding, H. P., Zhang, Y., Bertolotti, A., *et al.* (2000). Perk is essential for translational regulation and cell survival during the unfolded protein response, *Molecular Cell* **5**(5): 897–904.

Harper, M. E., Goddard, L., Glynne-Jones, E., *et al.* (2002). Multiple responses to EGF receptor activation and their abrogation by a specific EGF receptor tyrosine kinase inhibitor, *Prostate* **52**(1): 59–68.

Harris, A. M., Warner, B. W., Wilson, J. M., *et al.* (2007). Effect of α₁-adrenoceptor antagonist exposure on prostate cancer incidence: an observational cohort study, *J. Urol.* **178**: 2176–2180.

Harris, R. E., Beebe-Donk, J., Doss, H., *et al.* (2005). Aspirin, ibuprofen, and other non-steroidal anti-inflammatory drugs in cancer prevention: a critical review of non-selective COX-2 blockade, *Oncol. Rep.* **13**: 559–583.

Harris, T. J., von Maltzahn, G., Derfus, A. M., *et al.* (2006). Proteolytic actuation of nanoparticle self-assembly, *Angew. Chem. Int. Ed. Engl.* **45**: 3161–3165.

Harris, W. P., Mostaghel, E. A., Nelson, P. S., *et al.* (2009). Androgen deprivation therapy: progress in understanding mechanisms of resistance and optimizing androgen depletion, *Nat. Clin. Pract. Urol.* **6**: 76–85.

Hart, C. A., Brown, M., Bagley, S., *et al.* (2005). Invasive characteristics of human prostatic epithelial cells: understanding the metastatic process, *Br. J. Cancer.* **92**(3): 503–5012.

Hauck, C. R., Sieg, D. J., Hsia, D. A., *et al.* (2001) Inhibition of focal adhesion kinase expression or activity disrupts epidermal growth factor-stimulated signaling promoting the migration of invasive human carcinoma cells, *Cancer Res.* **61**: 7079.

Hayashi, K., Matsuda, S., Machida, K., *et al.* (2001). Invasion activating caveolin-1 mutation in human scirrhous breast cancers, *Cancer Res.* **61**(6): 2361–2364.

Hayek, O. R., Shabsigh, A., Kaplan, S. A., *et al.* (1999). Castration induces acute vaso-constriction of blood vessels in the rat prostate concomitant with a reduction of prostatic nitric oxide synthase activity, *J. Urol.* **162**(4): 1527–1531.

Hayes, S. A., Zarnegar, M., Sharma, M., *et al.* (2001). SMAD3 represses androgen receptor-mediated transcription, *Cancer Res.* **61**(5): 2112–2118.

Hayward, S. W., Grossfeld, G. D., Tlsty, T. D., *et al.* (1998). Genetic and epigenetic influences in prostatic carcinogenesis (review), *Int. J. Oncol.* **13**(1): 35–47.

Heinlein, C. A., and Chang, C. (2002). The roles of androgen receptors and androgen-binding proteins in nongenomic androgen actions, *Mol. Endocrinol.* **16**(10): 2181–2187.

Heinlein, C. A., and Chang, C. (2004). Androgen receptor in prostate cancer, *Endocr. Rev.* **25**(2): 276–308.

Heisler, L. E., Evangelou, A., Lew, A. M., *et al.* (1997). Androgen-dependent cell cycle arrest and apoptotic death in PC-3 prostatic cell cultures expressing a full-length human androgen receptor, *Mol. Cell Endocrinol.* **126**: 59–73.

Helbig, G., Christopherson, K. W. II, Bhat-Nakshatri, P., *et al.* (2003). NF-kappaB promotes breast cancer cell migration and metastasis by inducing the expression of the chemokine receptor CXCR4, *J. Biol. Chem.* **278**(24): 21631–21638.

Helzlsouer, K. J., Huang, H. Y., Alberg, A. J., *et al.* (2000). Association between alpha-tocopherol, gamma-tocopherol, selenium, and subsequent prostate cancer, *J. Natl. Cancer Inst.* **92**: 2018–2023.

Hendricksen, P. J. M., Dits, N. F. J., Kokame, K., *et al.* (2006). Evolution of the androgen receptor during progression of prostate cancer, *Cancer Res.* **66**: 5012.

Henshall, S. M., Quinn, D. I., Soon Lee, C., *et al.* (2001). Altered expression of androgen receptor in the malignant epithelium and adjacent stroma is associated with early relapse in prostate cancer, *Cancer Res.* **61**(2): 423–427.

Herceberg, S., Galan, P., Prezlosi, P., *et al.* (2004). The SU.VI.MAX study: a randomized, placebo-controlled trial of the health effects of antioxidant vitamins and minerals, *Arch. Intern Med.* **164**: 2335–2342.

Hernandez, I., Maddison, L. A., Wei, Y., *et al.* (2003). Prostate-specific expression of p53(R172L) differentially regulates p21, BAX, and mdm2 to inhibit prostate cancer progression and prolong survival, *Mol. Cancer Res.* **1**(14): 1036–1047.

Hernes, E., Fossa, S. D., Berner, A., *et al.* (2004). Expression of the epidermal growth factor receptor family in prostate carcinoma before and during androgen-independence, *Br. J. Cancer* **90**(2): 449–454.

Hess-Wilson, J. K., Daly, H. K., Zagorski, W. A., *et al.* (2006). Quantitating therapeutic disruption of tumor blood flow with intravital video microscopy, *Cancer Res.* **66**: 11998–12008.

Hetz, C., Bernasconi, P., Fisher, J., *et al.* (2006). Proapoptotic BAX and BAK modulate the unfolded protein response by a direct interaction with IRE1alpha, *Science* **312** (5773): 572–576.

Hill, M. M., Adrain, C., and Martin, S. J. (2003). Portrait of a killer: the mitochondrial apoptosome emerges from the shadows, *Mol. Interv.* **3**(1): 19–26.

Hippocrates, "Instruments of Reduction" (or Mochlicon) (c. 400 BC). Translated by E.T. Withington, *On Wounds in the Head, In the Surgery, On Fractures, On Joints, Mochlicon*. Harvard University Press, Harvard, 1928: 484.

Hiratuska, S., Nakamura, K., Iwai, S., *et al.* (2002). MMP-9 induction by vascular endothelial growth facto receptor-1 is involved in lung specific metastasis, *Cancer Cell* **2**: 289–300.

Hiroashi, S. (1998). Inactivation of the E-cadherin-mediated cell adhesion system in human cancers, *Am. J. Pathol.* **153**: 333–339.

Hitomi, J., Katayama, T., Eguchi, Y., *et al.* (2004). Involvement of caspase-4 in endoplasmic reticulum stress-induced apoptosis and Abeta-induced cell death, *J. Cell Biol.* **165**: 347–356.

Hjelmeland, A. B., Hjelmeland, M. D., Sji, Q., *et al.* (2005). Loss of phosphatase and tensin homologue increases transforming growth factor-β mediated invasion with enhanced Smad3 transcriptional activity, *Cancer Res.* **65**: 11276–11281.

Hobisch, A., Eder, I. E., Putz, T., *et al.* (1998). Interleukin-6 regulates prostate-specific protein expression in prostate carcinoma cells by activation of the androgen receptor, *Cancer Res.* **58**(20): 4640–4645.

Hochachka, P. W., Rupert, J. L., Goldenberg, L., *et al.* (2002). Going malignant: the hypoxia-cancer connection in the prostate, *Bioessays* **24**(8): 749–757.

Hockenbery, D., Nunez, G., Milliman, C., *et al.* (1990). BCL-2 is an inner mitochondrial membrane protein that blocks programmed cell death, *Nature* **348**: 334–336.

Hogue, A., Chen, H., and Xu, X. (2008). Statin induces apoptosis and cell growth arrest in prostate cancer cells, *Cancer Epidem. Biom.* **17**: 88–94.

Holash, J., Davis, S., Papadopoulos, N., *et al.* (2002). VEGF-Trap: A VEGF blocker with potent antitumor effects, *Proc. Natl. Acad. Sci. USA* **99**: 11393–11398.

Hollien, J., and Weissman, J. S. (2006). Decay of endoplasmic reticulum-localized mRNAs during the unfolded protein response, *Science* **313**(5783): 104–107.

Holmgren, L., O'Reilly, M. S., and Folkman, J. (1995). Dormancy of micrometastases: balanced proliferation and apoptosis in the presence of angiogenesis suppression, *Nat. Med.* **1**(2): 149–153.

Hong, J., Zhang, G., Dong, F., *et al.* (2002). Insulin-like growth factor (IGF)-binding protein-3 mutants that do not bind IGF-I or IGF-II stimulate apoptosis in human prostate cancer cells, *J. Biol. Chem.* **277**(12): 10489–10497.

Hood, J. D., and Cheresh, D. A. (2002). Role of integrins in cell invasion and migration, *Nat. Rev. Cancer* **2**: 91–100.

Horbinski, C., Mojesky, C., and Kyprianou, N. (2010). Live free or die: tales of homeless cells in cancer, *Am. J. Pathol.* 177: 1044–1052.

Horiguchi, K., Shirakihara, T., Nakano, A., *et al.* (2009). Role of Ras signaling in the induction of snail by transforming growth factor-β, *J. Biol. Chem.* 284: 245–253.

Horoszewicz, J. S., Leong, S. S., Kawinski, E. *et al.* (1983). LNCaP model of human prostatic carcinoma, *Cancer Res.* 43: 1809–1818.

Horsman, M. R., and Siemann, D. W. (2006). Pathophysiologic effects of vascular-targeting agents and the implications for combination with conventional therapies, *Cancer Res.* 66: 11520–11539.

Houzelstein, D., Goncalves, I. R., Fadden, A. J., *et al.* (2004). Phylogenetic analysis of the vertebrate galectin family, *Mol. Biol. Evol.* 21(7): 1177–1187.

Hsing, A. W. (2001). Hormones and prostate cancer: what's next? *Epidemiol. Rev.* 23(1): 42–58.

Hsu, A. L., Ching, T. T., Wang, D. S., *et al.* (2000). The cyclooxygenase-2 inhibitor celecoxib induces apoptosis by blocking Akt activation in human prostate cancer cells independently of BCL-2, *J. Biol. Chem.* 275: 11397–11403.

Hu, H., Jiang, C., Schuster, T., *et al.* (2006a). Inorganic selenium senstitizes prostate cancer cells to TRAIL-induced apoptosis through superoxide/p53/BAX-mediated activation of mitochondrial pathway, *Mol. Cancer Ther.* 5: 1873.

Hu, L., Hofmann, J., Zaloudek, C., *et al.* (2002). Vascular endothelial growth factor immunoneutralization plus Paclitaxel markedly reduces tumor burden and ascites in athymic mouse model of ovarian cancer, *Am. J. Pathol.* 161: 1917–1924.

Hu, P., Han, Z., Couvillon, A. D., *et al.* (2006b). Autocrine tumor necrosis factor alpha links endoplasmic reticulum stress to the membrane death receptor pathway through IRE1alpha-mediated NF-KB activation and down-regulation of TRAF2 expression, *Mol. Cell Biol.* 26: 3071–3084.

Hu, R., Dunn, T. A., Wei, S., *et al.* (2009). Ligand-independent androgen receptor variants derived from splicing of cryptic exons signify hormone-refractory prostate cancer, *Cancer Res.* 69: 16–22.

Huang, A., Gandour-Edwards, R., Rosenthal, S. A., *et al.* (1998). p53 and bcl-2 immuno-histochemical alterations in prostate cancer treated with radiation therapy, *Urology* 51(2): 346–351.

Huang, E. J., and Reichardt, L. F. (2003). Trk receptors: roles in neuronal signal trans-duction, *Annu. Rev. Biochem.* 72: 609–642.

Huang, S., Pettaway, C. A., Uehara, H., *et al.* (2001). Blockade of NF-kappaB activity in human prostate cancer cells is associated with suppression of angiogenesis, invasion and metastasis, *Oncogene* 20(31): 4188–4197.

Huber, M. A., Azoitei, N., Baumann, B., *et al.* (2004). NF-kappaB is essential for epithe-lial-mesenchymal transition and metastasis in a model of breast cancer progression, *J. Clin. Invest.* 114(4): 569–581.

Huggins, C. (1967). Endocrine-induced regression of cancers, *Cancer Res.* 27: 1925–1930.

Huggins, C., and Hodges, C. V. (1941). Studies on prostatic cancer. I. The effect of castration, of estrogen and androgen injection on serum phosphatases in metastatic carcinoma of the prostate, *Cancer Res.* 1: 293–297.

Huggins, C., Stevens, R. E., and Hodges, C. V. (1941). Studies on prostate cancer II. The effects of castration on advanced carcinoma of the prostate gland, *Arch. Surg.* **43**: 209–223.

Hugo, H., Ackland, M. L., Blick, T., *et al.* (2007). Epithelial-mesenchymal and mesenchymal-epithelial transitions in carcinoma progression, *J. Cell Physiol.* **213**: 374–383.

Humphrey, P. A. (2004). Gleason grading and prognostic factors in carcinoma of the prostate, *Modern Pathol.* **17**: 292–306.

Huss, W. J., Barrios, R. J., and Greenberg, N. M. (2003). SU5416 selectively impairs angiogenesis to induce prostate cancer-specific apoptosis, *Mol. Cancer Ther.* **2**: 611–616.

Hussain, A., Dawson N., Amin P., *et al.* (2005). Docetaxel followed by hormone therapy in men experiencing increasing prostate-specific antigen after primary local treatments for prostate cancer, *J. Clin. Oncol.* **23**: 2789–2796.

Hyer, M. L., Sudarshan, S., Kim, Y., *et al.* (2002). Downregulation of c-FLIP sensitizes DU145 prostate cancer cells to FAS-mediated apoptosis, *Cancer Biol. Ther.* **1(4)**: 401–406.

Hynes, R. O., and Lander. A. D. (1992). Contact and adhesive specificities in the associations, migrations and targeting of cells and axons, *Cell* **68**: 303–322.

Ibrahim, S. N., Lightner, V. A., Ventimiglia, J. B., *et al.* (1993). Tenascin expression in prostatic hyperplasia, intraepithelial neoplasia, and carcinoma, *Hum. Pathol.* **24(9)**: 982–989.

Imperato-McGinley, J., Binienda, Z., Arthur, A., *et al.* (1985). The development of a male pseudohermaphroditic rat using an inhibitor of the enzyme 5 alpha-reductase, *Endocrinology* **116(2)**: 807–812.

Ip, Y. T., Park, R. E., Kosman, D., *et al.* (1992). Dorsal-twist interactions establish snail expression in the presumptive mesoderm of the *Drosophila* embryo, *Genes Dev.* **6(8)**: 1518–1530.

Isaacs, J. T. (1984). The timing of androgen ablation therapy and/or chemotherapy in the treatment of prostatic cancer, *Prostate* **5(1)**: 1–17.

Isaacs, J. T. (1987). Response of glandular versus basal rat ventral prostatic epithelial cells to androgen withdrawal and replacement, *Prostate* **11(3)**: 229.

Isaacs, J. T., Lundmo, P. I., Berges, R., *et al.* (1992). Androgen regulation of programmed death of normal and malignant prostatic cells, *J. Androl.* **13(6)**: 457–464.

Ismail, H. A., Lessard, L., Mes-Masson, A. M., *et al.* (2004). Expression of NF-kappaB in prostate cancer lymph node metastases, *Prostate* **58(3)**: 308–313.

Itoh, S., Landstrom, M., Hermansson, A., *et al.* (1998). Transforming growth factor β1 induces nuclear export of inhibitory Smad7, *J. Biol. Chem.* **273**: 29195–29201.

Iwakoshi, N.N., Lee, A.H., Vallabhajosyula, P., *et al.* (2003). Plasma cell differentiation and the unfolded protein response intersect at the transcription factor XBP-1, *Nat. Immunol.* **4(4)**: 321–9.

Iyer, S., Wang, Z. G., Akhtari, M., *et al.* (2005). Targeting TGFb Signaling for Cancer Therapy, *Cancer Biol. Ther.* **4**: 261–266.

Jackson, M. W., Bentel, J. M., and Tilley, W. D. (1997). Vascular endothelial growth factor (VEGF) expression in prostate cancer and benign prostatic hyperplasia, *J. Urology* **157**: 2323–2328.

Jacobs, E. J., Rodriguez, C., Mondull, A. M., *et al.* (2005). A large cohort study of aspirin and other nonsteroidal anti-inflammatory drugs and prostate cancer, *J. Nat. Cancer Inst.* **97**: 975–980.

Jacobs, E. J., Thun, M. J., Bain, E. B., *et al.* (2007). A large cohort study of long-term daily use of adult-strength aspirin and cancer incidence, *J. Natl. Cancer Inst.* **99**: 608–615.

Jacobs, S. C. (1983). Spread of prostatic cancer to bone, *Urology* 21(4): 337–344.

Jacobsen, S. J., Girman, C. J., and Lieber, M. M. (2001). Natural history of benign prostatic hyperplasia, *Urology* 58(6S1): 5–16.

Jakowlew, S. B. (2006). Transforming growth factor-β in cancer and metastasis, *Cancer Metast. Rev.* 25(3): 435–457.

Janda, E., Lehmann, K., Killisch, I., *et al.* (2002). Ras and TGF[beta] cooperatively regulate epithelial cell plasticity and metastasis: dissection of Ras signaling pathways, *J. Cell Biol.* **156**(2): 299–313.

Jang, M., Park, B. C., Lee, A. Y., *et al.* (2007). Caspase-7 mediated cleavage of proteasome subunits during apoptosis, *Biochem. Biophys. Res. Commun.* **363**: 388–394.

Jarosch, E., Lenk, U., and Sommer, T. (2003). Endoplasmic reticulum-associated protein degradation, *Int. Rev. Cytol.* **223**: 39–81.

Jenster, G., van der Korput, H. A., van Vroonhoven, C., *et al.* (1991). Domains of the human androgen receptor involved in steroid binding, transcriptional activation and subcellular localization, *Mol. Endocrinol.* **5**: 1396–1404.

Jiang, H., Wek, S. A., McGrath, B. C., *et al.* (2003). Phosphorylation of the alpha subunit of eukaryotic initiation factor 2 is required for activation of NF-KB in response to diverse cellular stresses, *Mol. Cell Biol.* **25**: 5651–5663.

Jiang, Y., and Muschel, R. (2002). Regulation of matrix metalloproteinase-9 (MMP-9) by translational efficiency in murine prostate cancer cells, *Cancer Res.* **62**: 1910–1914.

Jiao, W., Datta, J., Lin, H. M., *et al.* (2006). Nucleocytoplasmic shuttling of the retinoblastoma tumor suppressor protein via Cdk phosphorylation-dependent nuclear export, *J. Biol. Chem.* **281**: 38098–38108.

Jin, B., Turner, L., Walters, W. A., *et al.* (1996). The effects of chronic high dose androgen or estrogen treatment on the human prostate [corrected], *J. Clin. Endocrinol. Metab.* **81**(12): 4290–4295.

Jin, R. J., Kwak, C., Lee, S. G., *et al.* (2000). The application of an anti-angiogenic gene (thrombospondin-1) in the treatment of human prostate cancer xenografts, *Cancer Gene Ther.* 7(12): 1537–1542.

Johansson, A., Rudolfsson, S. H., Wikström, P., *et al.* (2005). Altered levels of angiopoietin 1 and tie 2 are associated with androgen-regulated vascular regression and growth in the ventral prostate in adult mice and rats, *Endocrinology* 146(8): 3463–3470.

Johnstone, R. W., Ruefli, A. A., and Lowe, S. W. (2002). Apoptosis: a link between cancer genetics and chemotherapy, *Cell* 108(2): 1563–1564.

Jones, E., Pu, H., and Kyprianou, N. (2008). Targeting TGF-β in prostate cancer: therapeutic possibilities during tumor progression, *Expert Opin. Ther. Tar.* 13(2): 227–234.

Joseph, I. B., and Isaacs, J. T. (1997). Potentiation of the antiangiogenic ability of linomide by androgen ablation involves down-regulation of vascular endothelial growth factor in human androgen-responsive prostatic cancers, *Cancer Res.* **57**(6): 1054.

Joseph, I. B., Nelson, J. B., Denmeade, S. R., *et al.* (1997). Androgens regulate vascular endothelial growth factor content in normal and malignant prostatic tissue, *Clin. Cancer Res.* **3**(12:1): 2507–2511.

Joshi, B., Li, L., Taffe, B. G., *et al.* (1999). Apoptosis induction by a novel anti-prostate cancer compound, BMD188 (a fatty acid-containing hydroxamic acid), requires the mitochondrial respiratory chain, *Cancer Res.* **59**: 4343.

Jousse, C., Oyadomari, S., Novoa, I., *et al.* (2003). Inhibition of a constitutive translation initiation factor 2alpha phosphatase, CReP, promotes survival of stressed cells, *J. Cell Biol.* **163**(4): 767–775.

Jung, K., Lein, M., Ulbrich, N., *et al.* (1998). Quantification of matrix metalloproteinases and tissue inhibitors of metalloproteinase in prostatic tissue: analytical aspects, *Prostate* **34**(2): 130–136.

Kajiwara, T., Takeuchi, T., Ueki, T., *et al.* (1999). Effect of bcl-2 overexpression in human prostate cancer cells *in vitro* and *in vivo*, *Int. J. Urol.* **6**: 520–525.

Kalus, W., Zweckstetter, M., Renner, C., *et al.* (1998). Structure of the IGF-binding domain of the insulin-like growth factor-binding protein-5 (IGFBP-5): implications for IGF and IGF-I receptor interactions, *EMBO J.* **17**(22): 6558–6572.

Kamada, S., Shimono, A., Shinto, Y., *et al.* (1995). BCL-2 deficiency in mice leads to pleiotropic abnormalities: accelerated lymphoid cell death in thymus and spleen, polycystic kidney, hair hypopigmentation, and distorted small intestine, *Cancer Res.* **55**: 354–359.

Kamijo, T., Sato, T., Nagatomi, Y., *et al.* (2001). Induction of apoptosis by cyclooxygenase-2 inhibitors in prostate cancer cell lines, *Int. J. Urol.* **8**: S35–S39.

Kaminska, B., Wesolowska, A., and Danilkiewicz, M. (2005). TGF beta signalling and its role in tumour pathogenesis, *Acta Biochimica Polonica* **52**(2): 329–337.

Kamioka, H., and Yamashiro, T. (2008). Osteocytes and mechanical stress, *Clin. Calcium* **18**: 1287–1293.

Kane, L. P., Shapiro, V. S., Stokoe, D., *et al.* (1999). Induction of NF-kappaB by the Akt/PKB kinase, *Curr. Biol.* **9**(11): 601–604.

Kang, H. Y., Huang, K. E., Chang, S. Y., *et al.* (2002). Differential modulation of androgen receptor-mediated transactivation by Smad3 and tumor suppressor Smad4, *J. Biol. Chem.* **277**(46): 43749–43756.

Kang, H. Y., Lin, H. K., Hu, Y. C., *et al.* (2001). From transforming growth factor-β signaling to androgen action: identification of Smad3 as an androgen receptor coregulator in prostate cancer cells, *Proc. Natl. Acad. Sci. USA* **98**(6): 3018–3023.

Karin, M. (2006). Nuclear factor-κB in cancer development and progression, *Nature* **441**: 431–436.

Karin, M., Cao, Y., Greten, F. R., *et al.* (2002). NF-kappaB in cancer: from innocent bystander to major culprit, *Nat. Rev. Cancer* **2**(4): 301–310.

Kattan, M. W., Shariat, S. F., Andrews, B., *et al.* (2003). The addition of interleukin-6 soluble receptor and transforming growth factor beta1 improves a preoperative nomogram for predicting biochemical progression in patients with clinically localized prostate cancer, *J. Clin. Oncol.* **21(19)**: 3573–3579.

Kaufman, R. J. (1999). Stress signaling from the lumen of the endoplasmic reticulum: coordination of gene transcriptional and translational controls, *Genes Dev.* **13(10)**: 1211–1233.

Keledjian, K., Garrison, J. B., and Kyprianou, N. (2005). Doxazosin inhibits human vascular endothelial cell adhesion, migration, and invasion, *J. Cell Biochem.* **94(2)**: 374–388.

Keledjian, K., and Kyprianou, N. (2003). Anoikis induction by quinazoline based a1-adrenoceptor antagonists in prostate cancer cells: antagonistic effect of bcl-2, *J. Urol.* **169**: 1150–1156.

Kelliher, M. A., Grimm, S., Ishida, Y., *et al.* (1998). The death domain kinase RIP mediates the TNF- induced NF-κB signal, *Immunity* **8(3)**: 297–303.

Kerr, J. F., and Searle, J. (1973). Deletion of cells by apoptosis during castration-induced involution of the rat prostate, *Virchows Arch. B Cell Pathol.* **13(2)**: 87–102.

Kerr, J. F., Wyllie, A. H., and Currie, A. R. (1972). Apoptosis: a basic biological phenomenon with wide-ranging implications in tissue kinetics, *Br. J. Cancer* **26(4)**: 239.

Kim, C. S., Vasko, V. V., Kato, Y., *et al.* (2005). AKT activation promotes metastasis in a mouse model of follicular thyroid carcinoma, *Endocrinology* **146(10)**: 4456–4463.

Kim, D., Gregory, C. W., French, F. S., *et al.* (2002). Androgen receptor expression and cellular proliferation during transition from androgen-dependent to recurrent growth after castration in the CWR22 prostate cancer xenograft, *Am. J. Pathol.* **160**: 219–226.

Kim, H. E., Jiang, X., Du, F., *et al.* (2008). PHAPI, CAS, and Hsp70 promote apoptosome formation by preventing Apaf-1 aggregation and enhancing nucleotide exchange on Apaf-1, *Mol. Cell* **30(2)**: 239–247.

Kim, H. R., Lin, H. M., Biliran, H., *et al.* (1999). Cell cycle arrest and inhibition of anoikis by galectin-3 in human breast epithelial cells, *Cancer Res.* **59(16)**: 4148–4154.

Kim, O., Jiang, T., Xie, Y., *et al.* (2004). Synergism of cytoplasmic kinases in IL6-induced ligand-independent activation of androgen receptor in prostate cancer cells, *Oncogene* **23(10)**: 1838–1844.

Kim, S. J., Uehara, H., Karashima, T., *et al.* (2003). Blockade of epidermal growth factor receptor signaling in tumor cells and tumor-associated endothelial cells for therapy of androgen-independent human prostate cancer growing in the bone of nude mice, *Clin. Cancer Res.* **9(3)**: 1200–1210.

Kim, Y. M., Hwang, S., Kim, Y. M., *et al.* (2002). Endostatin blocks vascular endothelial growth factor-mediated signaling via direct interaction with KDR/Flk-1, *J. Biol. Chem.* **277(31)**: 27872–27879.

Kim, Y. M., Jang, J. W., Lee, O. H., *et al.* (2000). Endostatin inhibits endothelial and tumor cellular invasion by blocking the activation and catalytic activity of matrix metalloproteinase, *Cancer Res.* **60(19)**: 5410–5413.

Kimura, K., Bowen, C., Spiegel, S., *et al.* (1999). Tumor necrosis factor-alpha sensitizes prostate cancer cells to gamma-irradiation-induced apoptosis, *Cancer Res.* **59(7)**: 1606–1614.

Kinzler, K., and Vogelstein, B. (1998). Landscaping the cancer terrain, *Science* **280**: 1036–1037.

Kirsh, V. A., Hayes, R. B., Mayne, S. T., *et al.* (2006). PLCO trial, supplemental and dietary vitamin E, beta-carotene and vitamin C intakes and prostate cancer risk, *J. Natl. Cancer Inst.* **98**: 245–254.

Kischkel, F., Helbardt, S., Behrmann, I., *et al.* (1995). Cytotoxicity-dependent APO-1 (FAS/CD95)-associated proteins from a death-inducing signaling complex (DISC) with the receptor, *EMBO J.* **14**: 5579–5588.

Kischkel, F., Lawrence, D. A., Chuntharapai, A., *et al.* (2000). Apo2L/TRAIL-dependent recruitment of endogenous FADD and caspase-8 to death receptors 4 and 5, *Immunity* **12**: 611–620.

Kischkel, F., Lawrence, D. A., Chuntharapai, A., *et al.* (2001). Death receptor recruitment of endogenous caspase-10 and apoptosis initiation in the absence of caspase-8, *J. Biol. Chem.* **276**: 46639–46646.

Kisselev, A. F., and Goldberg, A. L. (2001). Proteasome inhibitors: from research tools to drug candidates, *Chem. Biol.* **8**: 739–758.

Klein, S., Giancotti, F. G., Presta, M., *et al.* (1993). Basic fibroblast growth factor modulates integrin expression in microvascular endothelial cells, *Mol. Biol. Cell* **4**: 973–982.

Kleiner, D., and Stetler-Stevenson, W. (1999). Matrix metalloproteinases and metastasis, *Cancer Chemother. Pharmacol.* **43**: S42–S51.

Klement, G., Baruchel, S., Rak, J., *et al.* (2000). Continuous low-dose therapy with vinblastine and VEGF receptor-2 antibody induces sustained tumor regression without overt toxicity, *J. Clin. Invest.* **105**: R15–R24.

Knudsen, K. E., and Scher, H. I. (2009). Starving the addiction: new opportunities for durable suppression of AR signaling in prostate cancer, *Clin. Cancer Res.* **15(15)**: 4792–4798.

Kojima, E., Takeuchi, A., Haneda, M., *et al.* (2003). The function of GADD34 is a recovery from a shutoff of protein synthesis induced by ER stress: elucidation by GADD34-deficient mice, *FASEB J.* **17(11)**: 1573–1575.

Konig, J. E., Senge, T., Allhoff, E. P., *et al.* (2004). Analysis of the inflammatory network in benign prostate hyperplasia and prostate cancer, *Prostate* **58(2)**: 121–129.

Kopetz, E. S., Nelson, J. B., and Carducci, M. A. (2002). Endothelin-1 as a target for therapeutic intervention in prostate cancer, *Invest. New Drugs* **20**: 173–182.

Korsmeyer, S. J., Wei, M. C., Saito, M., *et al.* (2000). Pro-apoptotic cascade activates BID, which oligomerizes BAK or BAX into pores that result in the release of cytochrome c, *Cell Death Differ.* **12**: 1166–1173.

Kornberg, L. J., and Liberti, J. P. (1992). Stimulations of Nb2 mitogenesis by an analogue of human growth hormone (110-127), *Biochem Int.* **28(5)**: 873–879.

Kote-Jarai, Z., Easton D. F., Stanford, J. L. *et al.* (2008). Multiple novel prostate cancer predisposition loci confirmed by an international study: The PRACTICAL Consortium, *Cancer Epidem. Biomar. Prev.* **17**: 2052–2061.

Krajewska, M., Krajewski, S., Banares, S., *et al.* (2003). Elevated expression of inhibitors of apoptosis proteins in prostate cancer, *Clin. Cancer Res.* **9**: 4914–4925.

Krajewski, S., Bodrug, S., Krajewska, M., *et al.* (1995). Immunohistochemical analysis of Mcl-1 protein in human tissues. Differential regulation of Mcl-1 and BCL-2 protein production suggests a unique role for Mcl-1 in control of programmed cell death *in vivo*, *Am. J. Pathol.* **146(6)**: 1309–1319.

Kraus, L. A., Samuel, S. K., Schmid, S. M., *et al.* (2003). The mechanism of action of docetaxel (Taxotere) in xenograft models is not limited to bcl-2 phosphorylation, *Invest. New Drugs* 21: 259–268.

Kreisberg, J. I., Malik, S. N., Prihoda, T. J., *et al.* (2004). Phosphorylation of Akt (Ser473) is an excellent predictor of poor clinical outcome in prostate cancer, *Cancer Res.* **64(15)**: 5232–5236.

Kreitzer, G., Liao, G., and Gundersen, G. G. (1999). Detyrosination of tubulin regulates the interaction of intermediate filaments with microtubules *in vivo* via a kinesin-dependent mechanism, *Mol. Biol. Cell* 10: 1105–1118.

Kroemer, G. (1997). The proto-oncogene BCL-2 and its role in regulating apoptosis, *Nat. Med.* 3: 614–620.

Krueckl, S. L., Sikes, R. A., Edlund, N. M., *et al.* (2004). Increased insulin-like growth factor I receptor expression and signaling are components of androgen-independent progression in a lineage-derived prostate cancer progression model, *Cancer Res.* **64(23)**: 8620–8629.

Kucharczak, J., Simmons, M. J., Fan, Y., *et al.* (2003). To be, or not to be: NF-kappaB is the answer-role of Rel/NF-kappaB in the regulation of apoptosis, *Oncogene* 22(56): 8961–8982.

Kuhajda, F. P. (2006). Fatty acid synthase and cancer: new application of an old pathway, *Cancer Res.* **66(12)**: 5977–5980.

Kuida, K., Haydar, T. F., Kuan, C-Y., *et al.* (1998). Reduced apoptosis and cytochrome c-mediated caspase activation in mice lacking caspase 9, *Cell* 94(3): 325–337.

Kulik, G., Carson, J. P., and Vomastek, T. (2001). Tumor necrosis factor alpha induces BID cleavage and bypasses antiapoptotic signals in prostate cancer LNCaP cells, *Cancer Res.* 61: 2713–2719.

Kulp, S. K., Chen, C. S., Wang, D. S., *et al.* (2006). Antitumor effects of a novel phenyl-butyrate-based histone deacetylase inhibitor, (S)-HDAC-42 in prostate cancer, *Clin. Cancer Res.* 12: 5199–51206.

Kumar, R., Srinivasan, S., Koduru, S., *et al.* (2009). Psoralidin, an herbal molecule, inhibits phosphatidylinositol 3-kinase-mediated Akt signaling in androgen-independent prostate cancer cells, *Cancer Prev. Res.* 2(3): 234–243.

Kurenova, E., Xu, L. H., Yang, X., *et al.* (2004). Focal adhesion kinase suppresses apoptosis by binding to the death domain of receptor-interacting protein, *Mol. Cell Biol.* **24(10)**: 4361–71.

Kurita, T., Wang, Y. Z., Donjacour, A. A., *et al.* (2001). Paracrine regulation of apoptosis by steroid hormones in the male and female reproductive system, *Cell Death Differ.* **8(2)**: 192–200.

Kuwana, T., Smith, J. J., Muzio, M., *et al.* (1998). Apoptosis induction by Caspase-8 is amplified through the mitochondrial release of cytochrome C, *Journal of Biological Chemistry* 273: 16589–16594.

Kwok, W. K., Ling, M. T., Lee, T. W., *et al.* (2005). Up-regulation of TWIST in prostate cancer and its implication as a therapeutic target, *Cancer Res.* **65(12)**: 5153–5162.

Kyprianou, N. (2003). Doxazosin and terazosin suppress prostate growth by inducing apoptosis: clinical significance, *J. Urol.* **169**: 1520–1525.

Kyprianou, N., Bruckheimer, E. M., and Guo, Y. (2000). Cell proliferation and apoptosis in prostate cancer: significance in disease progression and therapy, *Histolo. Histopathol.* **15**: 1211–1223.

Kyprianou, N., English, H. F., and Isaacs, J. T. (1988). Activation of a $Ca^{2+}$-$Mg^{2+}$-dependent endonuclease as an early event in castration-induced prostatic cell death, *Prostate* **13(2)**: 103.

Kyprianou, N., English, H. F., and Isaacs, J. T. (1990). Programmed cell death during regression of PC-82 human prostate cancer following androgen ablation, *Cancer Res.* **50(12)**: 3748.

Kyprianou, N., and Isaacs, J. T. (1988a). Activation of programmed cell death in the rat ventral prostate after castration, *Endocrinology* **122(2)**: 552–562.

Kyprianou, N., and Isaacs, J. T. (1988b). Identification of a cellular receptor for transforming growth factor-beta in the rat ventral prostate and its negative regulation by androgens, *Endocrinology* **123**: 2124–2131.

Kyprianou, N., and Isaacs, J. T. (1989). "Thymineless" death in androgen-independent prostatic cancer cells, *Biochem. Biophys. Res. Commun.* **165(1)**: 73.

Kyprianou, N., King, E. D., Bradbury, D., *et al.* (1997). BCL-2 over-expression delays radiation-induced apoptosis without affecting clonogenic survival in human prostate cancer cells, *Int. J. Cancer* **70**: 341–348.

Kyprianou, N., Tu, H., and Jacobs, S. C. (1996). Apoptotic and proliferative activities in human benign hyperplasia, *Hum. Pathol.* **27**: 668–675.

Lahm, H., Andre, S., Hoeflich, A., *et al.* (2004). Tumor galectinology: insights into the complex network of a family of endogenous lectins, *Glycoconj. J.* **20(4)**: 227–238.

Lakhani, S. A., Masud, A., Kuida, K., *et al.* (2006). Caspases 3 and 7: key mediators of mitochondrial events of apoptosis. *Science* **311**: 847–851.

Lamb, D. J., Weigel, N. L., and Marcelli, M. (2001). Androgen receptors and their biology, *Vitam. Horm.* **62**: 199–230.

Lara, P. N. Jr, Przemyslaw, T., and Quinn, D. L. (2004). Angiogenesis-targeted therapies in prostate cancer. *Clin. Prost. Cancer* **3**: 165–173.

Latha, K., Zhang, W., Cella, N., *et al.* (2005). Maspin mediates increased tumor cell apoptosis upon induction of the mitochondrial permeability transition, *Mol. Cell Biol.* **25(5)**: 1737–1748.

Lavery, D. N., and McEwan, I. J. (2008). Structural characterization of the native NH2-terminal transactivation domain of the human androgen receptor: a collapsed disordered conformation underlies structural plasticity and protein-induced folding, *Biochemistry* **47**: 3360–3369.

Lawler, J., Miao, W. M., Duquette, M., *et al.* (2001). Thrombospondin-1 gene expression affects survival and tumor spectrum of p53-deficient mice, *Am. J. Pathol.* **159(5)**: 1949–1956.

LeBlanc, R., Catley, L. P., Hideshima, T., *et al.* (2002). Proteasome inhibitor PS-341 inhibits human myeloma cell growth *in vivo* and prolongs survival in a murine model, *Cancer Res.* **62**: 4996–5000.

Lee, J. M., and Bernstein, A. (1993). p53 mutations increase resistance to ionizing radiation, *Proc. Natl. Acad. Sci.* **90**: 5742–5746.

Lee, K., Tirasophon, W., Shen, X., *et al.* (2002a). IRE1-mediated unconventional mRNA splicing and S2P-mediated ATF6 cleavage merge to regulate XBP1 in signaling the unfolded protein response, *Genes Dev.* 16(4): 452–466.

Lee, S. J., Jang, J. W., Kim, Y. M., *et al.* (2002b). Endostatin binds to the catalytic domain of matrix metalloproteinase-2, *FEBS Lett.* 519(1–3): 147–152.

Lee, S. O., Lou, W., Johnson, C. S., *et al.* (2004). Interleukin-6 protects LNCaP cells from apoptosis induced by androgen deprivation through the Stat3 pathway, *Prostate* 60(3): 178–186.

Leite, K. R., Franco, M. F., Srougi, M., *et al.* (2001). Abnormal expression of MDM2 in prostate carcinoma, *Mod. Pathol.* 14(5): 428–436.

Lekas, E., Johansson, M., Widmark, A., *et al.* (1997). Decrement of blood flow precedes the involution of the ventral prostate in the rat after castration, *Urol. Res.* 25(5): 309–314.

Leonard, G. D., Dahut, W. L., Gulley, J. L., *et al.* (2003). Docetaxel and thalidomide as a treatment option for androgen-independent, nonmetastatic prostate cancer, *Rev. Urol.* 3: S65–S70.

Léotoing, L., Manin, M., Monté, D., *et al.* (2007). Crosstalk between androgen receptor and epidermal growth factor receptor-signalling pathways: a molecular switch for epithelial cell differentiation, *J. Mol. Endocrinol.* 39(2): 151–162.

Levine, B., and Klionsky, D. J. (2004). Development by self-digestion: molecular mechanisms and biological functions of autophagy, *Dev. Cell* 6: 463–477.

Levine, B., and Kroemer, G. (2008). Autophagy in the pathogenesis of disease, *Cell* 132: 27–42.

Levine, A. C., Liu, X. H., Pietra, D., *et al.* (1998). Androgens induce the expression of vascular endothelial growth factor in human fetal prostatic fibroblasts, *Endocrinology* 139(11): 4672–4678.

Li, H., Stampfer, M. J., Giovannucci, E. L., *et al.* (2004). A prospective study of plasma selenium levels and prostate cancer risk, *J. Natl. Cancer Inst.* 96: 696–703.

Li, L., Ren, C. H., Tahir, S. A., *et al.* (2003a). Caveolin-1 maintains activated Akt in prostate cancer cells through scaffolding domain binding site interactions with and inhibition of serine/threonine protein phosphatases PP1 and PP2A, *Mol. Cell Biol.* 23(24): 9389–9404.

Li, L., Yang, G., Ebara, S., *et al.* (2001). Caveolin-1 mediates testosterone-stimulated survival/clonal growth and promotes metastatic activities in prostate cancer cells, *Cancer Res.* 61(11): 4386–4392.

Li, L., Yu, H., Schumacher, F., *et al.* (2003b). Relation of serum insulin-like growth factor-I (IGF-I) and IGF binding protein-3 to risk of prostate cancer (United States), *Cancer Cause. Control* 14(8): 721–726.

Li, Z., Shi, H. Y., and Zhang, M. (2005). Targeted expression of maspin in tumor vasculatures induces endothelial cell apoptosis, *Oncogene* 24(12): 2008–2019.

Liao, G., and Gundersen, G. G. (1998). Kinesin is a candidate for cross-bridging microtubules and intermediate filaments. Selective binding of kinesin to detyrosinated tubulin and vimentin, *J. Biol. Chem.* 273: 9797–9803.

Liao, Y., Grobholz, R., Abel, U., *et al.* (2003). Increase of AKT/PKB expression correlates with Gleason pattern in human prostate cancer, *Int. J. Cancer* 107(4): 676–680.

Lin, H. K., Yeh, S., Kang, H. Y., *et al.* (2001). Akt suppresses androgen-induced apoptosis by phosphorylating and inhibiting androgen receptor, *Proc. Natl. Acad. Sci. USA* **98**(13): 7200–7205.

Liotta, L. A., Steeg, P. S., and Stettler-Stevenson, W. G. (1991). Cancer metastasis and angiogenesis: an imbalance of positive and negative regulation, *Cell* **64**: 327–336.

Lippman, S. M., Klein, E. A., Goodman, P. J., *et al.* (2009). Effect of selenium and vitamin E on risk of prostate cancer and other cancers: the Selenium and Vitamin E Cancer Prevention Trial (SELECT), *JAMA* **301**: 39–51.

Lissbrant, I. F., Hammarsten, P., Lissbrant, E., *et al.* (2004). Neutralizing VEGF bioactivity with a soluble chimeric VEGF-receptor protein flt(1-3)IgG inhibits testosterone-stimulated prostate growth in castrated mice, *Prostate* **58**(1): 57–65.

Litvinov, I. V., De Marzo, A. M., and Isaacs, J. T. (2003). Is the Achilles' heel for prostate cancer therapy a gain of function in androgen receptor signaling? *J. Clin. Endocrinol. Metab.* **88**(7): 2972–2982.

Litvinov, I. V., Vander Griend, D. J., Antony, L., *et al.* (2006). Androgen receptor as a licensing factor for DNA replication in androgen-sensitive prostate cancer cells. *Proc. Natl. Acad. Sci. USA* **103**: 15085–15090.

Liu, J., Yin, S., Reddy, N., *et al.* (2004). Expand+BAX Mediates the Apoptosis-Sensitizing Effect of Maspin, *Cancer Res.* **64**: 1703.

Liu, X. H., Kirschenbaum, A., Yao, S., *et al.* (2000). Inhibition of cyclooxygenase-2 suppresses angiogenesis and the growth of prostate cancer *in vivo*, *J. Urol.* **164**: 820–825.

Liu, X. H., Kirschenbaum, A., Yu, K., *et al.* (2005a). Cyclooxygenase-2 suppresses hypoxia-induced apoptosis via a combination of direct and indirect inhibition of p53 activity in a human prostate cancer cell line, *J. Biol. Chem.* **280**(5): 3817–3823.

Liu, Y., Majumder, S., McCall, W., *et al.* (2005b). Inhibition of HER-2/neu kinase impairs androgen receptor recruitment to the androgen responsive enhancer, *Cancer Res.* **65**(8): 3404–3409.

Liu, Z-G., Hsu, H., Goeddel, D. V., and Karin, M. (1996). Dissection of TNF receptor 1 effector functions: JNK activation is not linked to apoptosis while NF-κB activation prevents cell death, *Cell* **87**(3): 565–576.

Logothetis, C. J., and Lin, S. H. (2005). Osteoblasts in prostate cancer metastasis to bone, *Nat. Rev. Cancer* **5**(1): 21–28.

Logothetis, C. J., Wu, K. K., Finn, L. D., *et al.* (2001). Phase I trial of the angiogenesis inhibitor TNP-470 for progressive androgen-independent prostate cancer, *Clin. Cancer Res.* **7**: 1198–1203.

Lomonaco, S. L., Finniss, S., Xiang C., *et al.* (2009). The induction of autophagy by gamma-radiation contributes to the radioresistance of glioma stem cells, *Int. J. Cancer* **125**: 717–722.

Lowsley, O. S. (1918). Surgical pathology of the human prostate gland, *Ann. Surg.* **68**(4): 399–415.

Loza-Coll, M. A., Perera, S., Shi, W., *et al.* (2005). A transient increase in the activity of Src-family kinases induced by cell detachment delays anoikis of intestinal epithelial cells, *Oncogene.* **24**: 1727–1737.

Lu, Y., Zhang, J., Dai, J., *et al.* (2004). Osteoblasts induce prostate cancer proliferation and PSA expression through interleukin-6-mediated activation of the androgen receptor, *Clin. Exp. Metastas.* 21: 399–408.

Lucia, M. S., Epstein, J. I., Goodman, P. J., *et al.* (2007). Finasteride and high-grade prostate cancer in the Prostate Cancer Prevention Trial, *J. Natl. Cancer Inst.* 99(18): 1375–1383.

Lu-Yao, G. L., Albertsen, P. C., Moore, D. F., *et al.* (2008). Survival following primary androgen deprivation therapy among men with localized prostate cancer, *JAMA* 300: 173–181.

Lyden, D., Hattori, K., Dias, S., *et al.* (2001). Impaired recruitment of bone-marrow-derived endothelial and hematopoietic precursor cells blocks tumor angiogenesis and growth, *Nat. Med.* 7: 1194–1201.

Lynch, D. K., Ellis, D.A., Edwards, P. A., *et al.* (1999). Integrin-linked kinase regulates phosphorylation of serine 473 of protein kinase B by an indirect mechanism, *Oncogene* 18: 8024–8032.

Lyons, J. F., Wilhelm, S., Hibner, B., *et al.* (2001). Discovery of a novel Raf kinase inhibitor, *Endocr. Relat. Cancer* 8: 219–225.

MacCorkle, R. A., Freeman, K. W., and Spencer, D. M. (1998). Synthetic activation of caspases: artificial death switches, *Proc. Natl. Acad. Sci. USA* 95(7): 3655–3660.

Murtola, T. J., Tammela, T. L. J., Lahtela, J., *et al.* (2007) Cholesterol-lowering drugs and prostate cancer risk: a population-based case-control study, *Cancer Epidem. Biomar.* 16: 2226–2232.

Mackey, T., Borkowski, A., Amin, P., *et al.* (1998). bcl-2/bax ratio as a predictive marker for therapeutic response to radiotherapy in patients with prostate cancer, *Urology* 52(6): 1085–1090.

Mahmud, S. M., Tanguay, S., Bégin, L. R., *et al.* (2006). Non-steroidal anti-inflammatory drug use and prostate cancer in a high-risk population, *Eur. J. Cancer Prev.* 15: 158–164.

Mailleux, A. A., Overholtzer, M., Schmelzle, T., *et al.* (2007). BIM regulates apoptosis during mammary ductal morphogenesis, and its absence reveals alternative cell death mechanism, *Dev. Cell* 12: 221–234.

Mainwaring, W. I. (1977). The mechanism of action of androgens, *Monogr. Endocrinol.* 10: 1.

Majumder, P. K., Febbo, P. G., Bikoff, R. *et al.* (2004). mTOR inhibition reverses Akt-dependent prostate intraepithelial neoplasia through regulation of apoptotic and HIF-1-dependent pathways, *Nat. Med.* 10: 594–601.

Makhsida, N., Shah, J., Yan, G., *et al.* (2005). Hypogonadism and metabolic syndrome: implications for testosterone therapy, *J. Urol.* 174: 827–834.

Malik, S. N., Brattain, M., Ghosh, P. M., *et al.* (2002). Immunohistochemical demonstration of phospho-Akt in high Gleason grade prostate cancer, *Clin. Cancer Res.* 8(4): 1168–1171.

Mancuso, A., Oudard, S., and Sternberg, C. N. (2007). Effective chemotherapy for hormone-refractory prostate cancer (HRPC): present status and perspectives with taxane-based treatments, *Crit. Rev. Oncol. Hematol.* 61: 176–185.

Mandell, L., Moran, A. P., Cocchiarella, A., *et al.* (2004). Intact gram-negative Helicobacter pylori, Helicobacter felis, and Helicobacter hepaticus bacteria activate innate immunity via toll-like receptor 2 but not toll-like receptor 4, *Infect. Immun.* 72(11): 6446–6454.

Manin, M., Baron, S., Goossens, K., *et al.* (2002). Androgen receptor expression is regulated by the phosphoinositide 3-kinase/Akt pathway in normal and tumoral epithelial cells, *Biochem. J.* 366(3): 729–736.

Manin, M., Martinez, A., Van Der Schuerens, B., *et al.* (2000). Acquisition of androgen-mediated expression of mouse vas deferens protein (MVDP) gene in cultured epithelial cells and in vas deferens during postnatal development, *J. Androl.* 21(5): 641–650.

Manin, M., Veyssiere, G., Cheyvialle, D., *et al.* (1992). *In vitro* androgenic induction of a major protein in epithelial cell subcultures from mouse vas deferens, *Endocrinology* 131(5): 2378–2386.

Manjeshwar, S., Branam, D. E., Lerner, M. R., *et al.* (2003). Tumor suppression by the prohibitin gene 3'untranslated region RNA in human breast cancer, *Cancer Res.* 63: 5251–5256.

Marcelli, M., Cunningham, G. R., Walkup, M., *et al.* (1999). Signaling pathway activated during apoptosis of the prostate cancer cell line LNCaP: overexpression of Caspase-7 as a new gene therapy strategy for prostate cancer, *Cancer Res.* 59: 382.

Marciniak, S. J., Yun, C. Y., Oyadomari, S., *et al.* (2004). CHOP induces death by promoting protein synthesis and oxidation in the stressed endoplasmic reticulum, *Genes Dev.* 18(24): 3066–3077.

Marks, L. S., Mostaghel, E. A., and Nelson, P. S. (2008). Prostate tissue androgens: history and current clinical relevance. *Urology*, 72: 375–380.

Marks, P. A., and Dokmanovic, M. (2005). Histone deacetylase inhibitors discovery and development as anticancer agents, *Expert Opin. Invest. Drugs* 14: 1497–1511.

Marks, P. A., Richon, V. M., and Rifkind, R. A. (1996). Cell cycle regulatory proteins are targets for induced differentiation of transformed cells: molecular and clinical studies employing hybrid polar compounds, *Int. J. Haemotol.* 63: 1–17.

Marneros, A. G., and Olsen, B. R. (2001). The role of collagen-derived proteolytic fragments in angiogenesis, *Matrix Biol.* 20(5–6): 337–345.

Martikainen, P., Kyprianou, N., Tucker, R. W., *et al.* (1991). Programmed death of nonproliferating androgen-independent prostatic cancer cells, *Cancer Res.* 51(17): 4693–4699.

Martin, S. S., Ridgeway, A. G., Pinkas, J., *et al.* (2004). A cytoskeleton-based functional genetic screen identifies Bcl-xL as an enhancer of metastasis, but not primary tumor growth, *Oncogene* 23: 4641–4645.

Massague, J., and Gomis, R. (2006). The logic of TGFβ signaling, *FEBS Letters* 580(12): 2811–2820.

Matthew, R., Karp, C. M., Beaudoin, B., *et al.* (2009). Autophagy suppresses tumorigenesis through elimination of p62, *Cell* 137: 1062–1075.

Matthews, E., Yang, T., Janulis, L., *et al.* (2000). Down-regulation of TGF-beta1 production restores immunogenicity in prostate cancer cells, *Br. J. Cancer* 83(4): 519–525.

Mawji, I. A., Simpson, C. D., Gronda, M., et al. (2007a). A chemical screen identifies anisomycin as an anoikis sensitizer that functions by decreasing FLIP protein synthesis, *Cancer Res.* **67**: 8307–8315.

Mawji, I. A., Simpson, C. D., Hurren, R., et al. (2007b). Critical role for Fas-associated death domain-like interleukin-1-converting enzyme-like inhibitory protein in anoikis resistance and distant tumor formation, *J. Natl. Cancer Inst.* **99**: 811–822.

Maynard, A. D., and Pui, D. Y. H. (2007). Nanotechnology and occupational health: new technologies – new challenges, *Journal of Nanoparticle Research* **9**: 1–3.

McArdle, P. A., Canna, K., McMillan, D. C., et al. (2004). The relationship between T-lymphocyte subset infiltration and survival in patients with prostate cancer, *Br. J. Cancer* **91**: 541–543.

McCarty, M. F. (2004). Targeting multiple signaling pathways as a strategy for managing prostate cancer: multifocal signal modulation therapy, *Integr. Cancer Ther.* **3**(4): 349–380.

McConnell, J. D., Roehrborn, C. G., Bautista, O. M., et al. (2003). The long-term effect of doxazosin, finasteride and combination therapy on the clinical progression of benign prostate hyperplasia, *N. Engl. J. Med.* **349**: 2387–2398.

McCullough, K. D., Martindale, J. L., Klotz, L. O., et al. (2001). Gadd153 sensitizes cells to endoplasmic reticulum stress by down-regulating Bcl2 and perturbing the cellular redox state, *Mol. Cell Biol.* **21**(4): 1249–1259.

McDonnell, T. J., Deane, N., Platt, F. M., et al. (1989). bcl-2-Immunoglobin transgenic mice demonstrate extended B cell survival and follicular lymphoproliferation, *Cell* **57**(1): 79–88.

McDonnell T. J., and Korsmeyer S. J. (1991). Progression from lymphoid hyperplasia to high-grade malignant lymphoma in mice transgenic for the t(14; 18), *Nature* **49**(6306): 254–256.

McDonnell, T. J., Navone, N. M., Troncoso, P., et al. (1997). Expression of bcl-2 onco-protein and p53 protein accumulation in bone marrow metastases of androgen independent prostate cancer, *J. Urol.* **157**: 569–574.

McDonnell, T. J., Troncoso, P., Brisbay, S. M., et al., (1992). Expression of the protoncogene bcl-2 in the prostate and its association with emergence of androgen-independent prostate cancer, *Cancer Res.* **52**: 6940–6944.

McEleny, K. R., Watson, R. W. G., and Fitzpatrick, J. M. (2001). Defining the role for the inhibitors of apoptosis proteins in prostate cancer, *Prost. Ca. Prostat. Dis.* **4**: 28–32.

McEntee, M. F., Ziegler, C., Reel, D., et al. (2008). Dietary n-3 polyunsaturated fatty acids enhance hormone ablation therapy in androgen-dependent prostate cancer, *Am. J. Pathol.* **173**(1): 256–268.

McKenzie, S., and Kyprianou, N. (2006). Apoptosis evasion: the role of survival pathways in prostate cancer progression and therapeutic resistance, *J. Cell Biochem.* **97**(1): 18.

McKenzie, S., Sakamoto, S., and Kyprianou, N. (2008). Maspin sensitizes apoptotic response of prostate cancer cells to hypoxia, *Oncogene* **27**: 7171–7179.

McMenamin, M. E., Soung, P., Perera, S., et al. (1999). Loss of PTEN expression in paraffin-embedded primary prostate cancer correlates with high Gleason score and advanced stage, *Cancer Res.* **59**: 4291–4296.

McNeal, J. E. (1978). Origin and evolution of benign prostatic enlargement, *Invest. Urol.* **15**: 340–345.

McNeal, J. E. (1992). Cancer volume and site of origin of adenocarcinoma in the prostate: relationship to local and distant spread, *Human Pathol.* 23: 258–266.

McNeal, J. E., Redwine, E. A., Freiha, F. S., *et al.* (1988). Zonal distribution of prostatic adenocarcinoma. correlation with histologic pattern and direction of spread, *Am. J. Surg. Pathol.* 12: 897–906.

Medema, J. P., Scaffidi, C., Kischkel, F. C., *et al.* (1997). FLICE is activated by association with the CD95 death-inducing signaling complex (DISC), *EMBO J.* 16(10): 2794–2804.

Mehlen, P., and Puisieux, A. (2006). Metastasis: a question of life and death, *Nat. Rev. Cancer* 6(6): 449–458.

Mellinghoff, I. K., Vivanco, I., Kwon, A., *et al.* (2004). HER2/neu kinase-dependent modulation of androgen receptor function through effects on DNA binding and stability, *Cancer Cell* 6(5): 517–527.

Memarzadeh, S., Xin, L., Mulholland, D. J., *et al.* (2007). Enhanced paracrine FGF10 expression promotes formation of multifocal prostate adenocarcinoma and an increase in epithelial androgen receptor, *Cancer Cell* 12(6): 572–585.

Mendel, D. B., Laird, A. D., Xin, X., *et al.* (2003). *In vivo* antitumor activity of SU11248, a novel tyrosine kinase inhibitor targeting vascular endothelial growth factor and platelet-derived growth factor receptors: determination of a pharmacokinetic/pharmacodynamic relationship, *Clin. Cancer Res.* 9: 327–337.

Michalaki, V., Syrigos, K., Charles, P., *et al.* (2004). Serum levels of IL-6 and TNF-α correlate with clinicopathological features and patient survival in patients with prostate cancer, *Br. J. Cancer* 90: 2312–2316.

Micheau, O., and Tschopp, J. (2003). Induction of TNF Receptor I- mediated apoptosis via two sequential signaling complexes, *Cell* 114(2): 181–190.

Migliaccio, A. G., Castoria, M., Di Domenico, A. *et al.* (2000). Steroid-induced androgen receptor-oestradiol receptor beta-Src complex triggers prostate cancer cell proliferation, *EMBO J.* 19: 5406–5417.

Millauer, B., Wizigmann-Voos, S., Schnürch, H., *et al.* (1993). High affinity VEGF binding and developmental expression suggests flk-1 as a major regulator of vasculogenesis and angiogenesis, *Cell* 72: 835–846.

Mimeault, M., Pommery, N., Henichart, J. P. (2003). New advances on prostate carcinogenesis and therapies: involvement of EGF-EGFR transduction system, *Growth Factors* 21(1): 1–14.

Minucci, S., and Pelicci, P. G. (2006). Histone deacetylase inhibitors and the promise of epigenetic (and more) treatments for cancer, *Nat. Rev. Cancer* 6: 38–51.

Miyake, H., Nelson, C., Rennie, P. S., *et al.* (2000). Overexpression of insulin-like growth factor binding protein-5 helps accelerate progression to androgen-independence in the human prostate LNCaP tumor model through activation of phosphatidylinositol 3'-kinase pathway, *Endocrinology* 141(6): 2257–2265.

Miyamoto, S., Yano, K., Sugimoto, S., *et al.* (2004). Matrix metalloproteinase-7 facilitates insulin-like growth factor bioavailability through its proteinase activity on insulin-like growth factor binding protein 3, *Cancer Res.* 64(2): 665–671.

Mizokami, A., Gotoh, A., Yamada, H., *et al.* (2000). Tumor necrosis factor-alpha represses androgen sensitivity in the LNCaP prostate cancer cell line, *J. Urol.* 164: 800–805.

Mohler, J. L., Greggory, C. W., Ford O. H., *et al.* (2004). The androgen axis in recurrent prostate cancer, *Clin. Cancer Res.* **10**: 440–448

Montgomery, R. B., Bonham, M., Nelson, P. S., *et al.* (2005). Estrogen effects on tubulin expression and taxane mediated cytotoxicity in prostate cancer cells, *Prostate* **65**: 141–150.

Montgomery, R. B., Mostaghel, E. A., Vessella, R., *et al.* (2008). Maintenance of intratumoral androgens in metastatic prostate cancer: mechanism for castration-resistant tumor growth, *Cancer Res.* **68**: 4447–4454.

Montpetit, M. L., Lawless, K. R., and Tenniswood, M. (1986). Androgen repressed messages in the rat ventral prostate, *Prostate* **8**: 25–36.

Mordenti, J., Thomsen, K., Licko, V., *et al.* (1999). Efficacy and concentration-response of murine anti-VEGF monoclonal antibody in tumor-bearing mice and extrapolation to humans, *Toxicol. Pathol.* **27**: 14–21.

Morton, D. M., Barrack, E. R. (1995). Modulation of transforming growth factor β1 effects on prostate cancer cell proliferation by growth factors and extracellular matrix, *Cancer Res* **55**: 2596–2602.

Moscat, J., and Diaz-Meco, M. T. (2009). p62 at the crossroads of autophagy, apoptosis, and cancer, *Cell* **137**: 1001–1004.

Moschos, S. J., and Mantzoros, C. S. (2002). The role of the IGF system in cancer: from basic to clinical studies and clinical applications, *Oncology* **63**(4): 317–332.

Mosquera, J. M., Perner, S., Genega, E. M., *et al.* (2008). Characterization of TMPRSS2-ERG Fusion high-grade prostatic intraepithelial neoplasia and potential clinical implications, *Clin. Cancer Res.* **14**: 3380–3385.

Mostaghel, E. A., Page, S. T., Lin, D. W., *et al.* (2007). Intraprostatic androgens and androgen-regulated gene expression persist after testosterone suppression: therapeutic implications for castration-resistant prostate cancer, *Cancer Res.* **67**(10): 5033–5041.

Moustakas, A., and Heldin, C-H. (2005). Non-Smad TGF-β signals, *J. Cell Sci.* **118**: 9573–9584.

Moustakas, A., Kowanetz, M., and Thuault, S. (2006). TGF-β/SMAD Signaling in Epithelial to Mesenchymal Transition, *Proteins Cell Reg.* **5**: 131–150.

Moutsopoulos, N. M., Wen, J., and Wahl, S. M. (2008). TGF-beta and tumors—an ill-fated alliance, *Curr. Opin. Immunol.* **20**: 234–240.

Movsas, B., Chapman, J. D., Greenberg, R. E., *et al.* (2000). Increasing levels of hypoxia in prostate carcinoma correlate significantly with increasing clinical stage and patient age: an Eppendorf pO(2) study, *Cancer* **89**(9): 2018–2024.

Muenchen, H. J., Lin, D. L., Walsh, M. A., *et al.* (2000). Tumor necrosis factor-alpha-induced apoptosis in prostate cancer cells through inhibition of nuclear factor-κB by an IκBα "super-repressor", *Clin. Cancer Res.* **6**(5): 1969–1977.

Murphy, W. M., Soloway, M. S., and Barrows, G. H. (1991). Pathologic changes associated with androgen deprivation therapy for prostate cancer, *Cancer* **68**(4): 821.

Mutsaers, S. E., McAnulty, R. J., Laurent, G. J., *et al.* (1997). Cytokine regulation of mesothelial cell proliferation *in vitro* and *in vivo*, *Eur. J. Cell Biol.* **72**(1): 24–29.

Muzio, M., Chinnaiyan, A. M., Kischkel, F. C., *et al.* (1996). FLICE, a novel FADD-homologous ICE/CED-3-like protease, is recruited to the CD95 (Fas/APO-1) death-inducing signaling complex, *Cell* **85**: 817–827.

Myers, R. B., Oelschlager, D., Manne, U., *et al.* (1999). Androgenic regulation of growth factor and growth factor receptor expression in the CWR22 model of prostatic adenocarcinoma, *Int. J. Cancer* **82**(3): 424–429.

Nakagawa, M., Oda, Y., Eguchi, T., *et al.* (2007). Expression profile of class I histone deacetylases in human cancer tissues, *Oncol. Rep.* **18**: 769–774.

Nakagawa, T., Zhu, H., Morishima, N., *et al.* (2000). Caspase-12 mediates endoplasmic-reticulum-specific apoptosis and cytotoxicity by amyloid-beta, *Nature* **403**(6765): 98.

Nakahara, S., Oka, N., and Raz, A. (2005). On the role of galectin-3 in cancer apoptosis, *Apoptosis* **10**(2): 267–275.

Nakai, Y., Nelson, W. G., and De Marzo, A. M. (2007). The dietary charred meat carcinogen 2-amino-1-methyl-6-phenylimidazo[4,5-b]pyridine acts as both a tumor initiator and promoter in the rat ventral prostate, *Cancer Res.* **67**: 1378–1384.

Nakano, K., Fukabori, Y., Itoh, N., *et al.* (1999). Androgen-stimulated human prostate epithelial growth mediated by stromal-derived fibroblast growth factor-10, *Endocr. J.* **46**(3): 405–413.

Nakayama, K., Negishi, I., Kuida, K., *et al.* (1994). Targeted disruption of BCL-2alpha-beta in mice: occurrence of gray hair, polycystic kidney disease, and lymphocytopenia, *Proc. Natl. Acad. Sci. USA* **91**: 3700–3704.

Nalepa, G., Rolfe, M., and Harper, J. W. (2006). Drug discovery in the ubiquitin-proteasome system, *Nat. Rev. Drug Discov.* **5**: 596–613.

Nam, S., Kim, D., Cheng, J. Q., *et al.* (2005). Action of the Src family kinase inhibitor, dasatinib (BMS-354825), on human prostate cancer cells, *Cancer Res.* **65**: 9185–9189.

Nangia-Makker, P., Honjo, Y., Sarvis, R., *et al.* (2000). Galectin-3 induces endothelial cell morphogenesis and angiogenesis, *Am. J. Pathol.* **156**(3): 899–909.

Narayanan, B. A., Narayanan, N. K., Pitman B., *et al.* (2006). Adenocarcina of the mouse prostate growth inhibition by celecoxib: downregulation of transcription factors involved in COX-2 inhibition, *Prostate* **66**: 257–265.

Narayanan, N. K., Nargi, D., Horton, L., *et al.* (2009). Inflammatory processes of prostate tissue microenvironment drive rat prostate carcinogenesis: preventive effects of celecoxib, *Prostate* **69**: 133–141.

Nasu, Y., Timme, T. L., Yang, G., *et al.* (1998). Suppression of caveolin expression induces androgen sensitivity in metastatic androgen-insensitive mouse prostate cancer cells, *Nat. Med.* **4**(9): 1062–1064.

Navone, N. M., Troncoso, P., Pisters, L. L., *et al.* (1993). p53 protein accumulation and gene mutation in the progression of human prostate carcinoma, *J. Natl. Cancer Inst.* **85**(20): 1657–1669.

Nelson, J. B., Hedican, S. P., George, D. J., *et al.* (1995). Identification of endothelin-1 in the pathophysiology of metastatic adenocarcinoma of the prostate, *Nature Med.* **1**(9): 944–949.

Nelson, W. G. (2007). Prostate cancer prevention, *Curr. Opin. Urol.* **17**: 157–167.

Nelson, W. G., De Marzo, A. M., DeWeese, T. L., *et al.* (2004). The role of inflammation in the pathogenesis of prostate cancer, *J. Urol.* **172**(5:2): S6–S11.

Newmeyer, D. D., and Miller, S. F. (2003). Mitochondria: releasing the power for life and unleashing the machineries of death, *Cell* **112**: 481–490.

Ng, S. S., MacPherson, G. R., Gutschow, M., *et al.* (2004). Antitumor effects of thalidomide analogs in human prostate cancer xenografts implanted in immunodeficient mice, *Clin. Cancer Res.* 10: 4192–4197.

Ni, J., Xingqiao, W., Yao, J., *et al.* (2005). Tocopherol-associated protein suppresses prostate cancer cell growth by inhibition of the phosphoinositide 3-kinase pathway, *Cancer Res.* 65(21): 9807–9815.

Nicholson, B., Gulding, K., Conaway, M., *et al.* (2004). Combination antiangiogenic and androgen deprivation therapy for prostate cancer: a promising therapeutic approach, *Clin. Cancer Res.* 10: 8728–8734.

Nickerson, T., Chang, F., Lorimer, D., *et al.* (2001). *In vivo* progression of LAPC-9 and LNCaP prostate cancer models to androgen independence is associated with increased expression of insulin-like growth factor I (IGF-I) and IGF-I receptor (IGF-IR), *Cancer Res.* 61(16): 6276–6280.

Nieto, M. A. (2002). The Snail superfamily of zinc-finger transcription factors, *Nat. Rev. Mol. Cell Biol.* 3(3): 155–166.

Nijtmans, L. G. J., de Jong, L., Sanz, M. A., *et al.* (2000). Prohibitin acts as a membrane-bound chaperone for the stabilization of mitochondrial proteins, *EMBO J.* 19: 2444–2451.

Nikitina, E. Y., Desai, S. A., Zhao, X., *et al.* (2005). Versatile prostate cancer treatment with inducible caspase and interleukin-12, *Cancer Res.* 65(10): 4309–19.

Nikolopoulos, S. N., Blaikie, P., Yoshioka, T., *et al.* (2004). Integrin beta4 signaling promotes tumor angiogenesis, *Cancer Cell* 6: 471–483.

Nishitoh, H., Matsuzawa, A., Tobiume, K., *et al.* (2002). ASK1 is essential for endoplasmic reticulum stress-induced neuronal cell death triggered by expanded polyglutamine repeats, *Genes Dev.* 16(11): 1345–1355.

Nishitoh, H., Saitoh, M., Mochida, Y., *et al.* (1998). ASK1 is essential for JNK/SAPK activation by TRAF2, *Mol. Cell* 2(3): 389–395.

Nishiyama, T., Hashimoto, Y., and Takahashi, K. (2004). The influence of androgen deprivation therapy on dihydrotestosterone levels in the prostatic tissue of patients with prostate cancer. *Clin. Cancer Res.* 10: 7121–7126.

Niu, Y., Xu, Y., Zhang, J., *et al.* (2001). Proliferation and differentiation of prostatic stromal cells, *BJU Int.* 87(4): 386–393.

Noh, Y. H., Matsuda, K., Hong, Y. K., *et al.* (2003). An N-terminal 80 kDa recombinant fragment of human thrombospondin-2 inhibits vascular endothelial growth factor induced endothelial cell migration *in vitro* and tumor growth and angiogenesis *in vivo*, *J. Invest. Dermatol.* 121(6): 1536–1543.

Novoa, I., Zhang, Y., Zeng, H., *et al.* (2003). Stress-induced gene expression requires programmed recovery from translational repression, *EMBO J.* 22(5): 1180–1187.

Nyberg, P., Heikkila, P., Sorsa, T., *et al.* (2003). Endostatin inhibits human tongue carcinoma cell invasion and intravasation and blocks the activation of matrix metalloprotease-2, -9, and -13, *J. Biol. Chem.* 278(25): 22404–22411.

Nyberg, P., Xie, L., and Kalluri, R. (2005). Endogenous inhibitors of angiogenesis, *Cancer Res.* 65(10): 3967–3979.

Obeng, E. A., and Boise, L. H. (2005). Caspase-12 and caspase-4 are not required for caspase-dependent endoplasmic reticulum stress-induced apoptosis, *J. Biol. Chem.* 280(33): 29578–29587.

Ohlson, N., Bergh, A., Stattin, P., *et al.* (2007). Castration-induced epithelial cell death in human prostate tissue is related to locally reduced IGF-1 levels, *Prostate* **67**(1): 32–40.

Oka, N., Takenaka, Y., and Raz, A. (2004). Galectins and urological cancer, *J. Cell Biochem.* **91**(1): 118–124.

Okada, H., and Mak, T. W. (2004). Pathways of apoptotic and non-apoptotic death in tumour cells, *Nat. Rev. Cancer* **4**(8): 592–603.

Oliver, C. L., Miranda, M. B., Shangary, S., *et al.* (2005). (-)-Gossypol acts directly on the mitochondria to overcome BCL-2- and Bcl-X(L)-mediated apoptosis resistance, *Mol. Cancer Ther.* **4**: 23–31.

Oltvai, Z. N., Milliman, C. L., and Korsmeyer, S. J. (1993). BCL-2 heterodimerizes *in vivo* with a conserved homolog, BAX, that accelerates programmed cell death, Cell **74**: 609–619.

Onder, T. T., Gupta, P. B., Mani, S. A., *et al.* (2008). Loss of E-cadherin promotes metastasis via multiple downstream transcriptional pathways, *Cancer Res.* **68**: 3645–3654.

Ono, S. (2003). Regulation of actin filament dynamics by actin depolymerizing factor/cofilin and actin-interacting protein 1: new blades for twisted filaments, *Biochemistry* **42**: 13363–13368.

Oosterhoff, J. K., Grootegoed, J. A., and Blok, L. J. (2005). Expression profiling of androgen-dependent and -independent LNCaP cells: EGF versus androgen signaling, *Endocr. Relat. Cancer* **12**: 135–148.

O'Reilly, M. S., Boehm, T., Shing, Y., *et al.* (1997). Endostatin: an endogenous inhibitor of angiogenesis and tumor growth, *Cell* **88**(2): 277–285.

Orimo, A., Gupta, P. B., Sgroi, D. C., *et al.* (2005). Stromal fibroblasts present in invasive human breast carcinomas promote tumor growth and angiogenesis through elevated SDF-1/CXCL12 secretion, *Cell* **121**(3): 335–348.

Orio, F. Jr, Terouanne, B., Georget, V., *et al.* (2002). Potential action of IGF-1 and EGF on androgen receptor nuclear transfer and transactivation in normal and cancer human prostate cell lines, *Mol. Cell Endocrinol.* **198**(1–2): 105–114.

Orlandi, M., Marconcini, L., Ferruzi, R., *et al.* (1996). Identification of a c-fos-induced gene that is related to the platelet-derived growth factor/vascular endothelial growth factor family, *Proc. Natl. Acad. Sci. USA* **93**: 11675–11680.

Orlowski, R. Z., and Baldwin, A. S. Jr. (2002). NF-kappaB as a therapeutic target in cancer, *Trends Mol. Med.* **8**(8): 385–389.

Ornitz, D. M., and Itoh, N. (2001). Fibroblast growth factors, *Genome Biol.* **2**(3): REVIEWS3005.

Orr-Urtreger, A., Bar-Shira, A., Matzkin, H., *et al.* (2007). The homozygous P582S mutation in the oxygen-dependent degradation domain of HIF-1α is associated with increased risk for prostate cancer, *Prostate* **67**: 8–13.

Ozdamar, B., Rose, R., Barrios-Rodiles, M., *et al.* (2005). Regulation of the polarity protein Par6 by TGFβ receptors controls epithelial cell plasticity, *Science* **307**: 1603–1609.

Ozes, O. N., Mayo, L. D., Gustin, J. A., *et al.* (1999). NF-kappaB activation by tumour necrosis factor requires the Akt serine-threonine kinase, *Nature* **401**: 82–85.

Page, S. T., Lin, D. W., Mostaghel. E. A., *et al.* (2006). Persistent intraprostatic androgen concentrations after medical castration in healthy men, *J. Clin. Endocr. Metab.* **91(10)**: 3850–3856.

Papandreou, C. N., Daliani, D. D., Nix, D., *et al.* (2004). Phase I trial of the proteasome inhibitor bortezomib in patients with advanced solid tumors with observations in androgen-independent prostate cancer, *J. Clin. Oncol.* **22**: 2108–2121.

Papandreou, C. N., and Logothetis, C. J. (2004). Bortezomib as a potential target for prostate cancer, *Cancer Res.* **64**: 5036–5043.

Park, J., Lee, M. G., Cho, K., *et al.* (2003). Transforming growth factor-β1 activates interleukin-6 expression in prostate cancer cells through the synergistic collaboration of the Smad2, p38-NF-κB, JNK, and Ras signaling pathways, *Oncogene* **22**: 4314–4332.

Park, S. I., Zhang, J., Phillips, K. A., *et al.* (2008). Targeting SRC family kinases inhibits growth and lymph node metastases of prostate cancer in an orthotopic nude mouse model, *Cancer Res.* **68**: 3323–3333.

Partin, J. V., Anglin, I. E., and Kyprianou, N. (2003). Quinazoline-based $\alpha_1$-adrenoceptor antagonists induce prostate cancer cell apoptosis via TGF-β signaling and IκBα induction, *Br. J. Cancer* **88(10)**: 1615–1621.

Pattingre, S., Espert, L., Biard-Piechaczyk, M., *et al.* (2008). Regulation of macroautophagy by mTOR and Beclin 1 complexes, *Biochimie* **90** 313–323.

Paweletz, C. P., Charboneau, L., Bichsel, V. E., *et al.* (2001). Reverse phase protein microarrays which capture disease progression show activation of pro-survival pathways at the cancer invasion front, *Oncogene* **20**: 1981–1989.

Pedram, A., Razandi, M., Sainson, R. C. A., *et al.* (2007). A conserved mechanism for steroid receptor translocation of the plasma membrane, *J. Biol. Chem.* **282**: 22278–22288.

Peinado, H., Olmeda, D., and Cano, A. (2007). Snail, Zeb and bHLH factors in tumour progression: an alliance against the epithelial phenotype? *Nat. Rev. Cancer* 7: 415–428.

Perez-Moreno, M., and Fuchs, E. (2006). Catenins: keeping cells from getting their signals crossed, *Dev. Cell* 11: 601–612.

Perlman, H., Zhang, X., Chen, M. W., *et al.* (1999). An elevated bax/bcl-2 ratio corresponds with the onset of prostate epithelial cell apoptosis, *Cell Death Differ.* **6(1)**: 48.

Persad, S., Attwell, S., Gray, V., *et al.* (2000). Inhibition of integrin-linked kinase (ILK) suppresses activation of protein kinase B/Akt and induces cell cycle arrest and apoptosis of PTEN-mutant prostate cancer cells. *Proc. Natl. Acad. Sci.* **97(7)**: 3207–3212.

Peter, M. E., and Krammer, P. H. (2003). The CD95(APO-1/Fas) DISC and beyond, *Cell Death Differ.* **10**: 26–35.

Peterziel, H., Mink, S., Schonert, A., *et al.* (1999). Rapid signalling by androgen receptor in prostate cancer cells, *Oncogene* **18(46)**: 6322–6329.

Petrylak, D. P. (2005). The current role of chemotherapy in metastatic hormone-refractory prostate cancer, *Urology* **65**: 3–8.

Petrylak D. P., Tangen C. M., Hussain M. H., *et al.* (2004) Docetaxel and estramustine compared with mitoxantrone and prednisone for advanced refractory prostate cancer. *N. Engl. J. Med.* **351**: 1513–20.

Pfeil, K., Eder, I. E., Putz, T., *et al.* (2004). Long-term androgen-ablation causes increased resistance to PI3K/Akt pathway inhibition in prostate cancer cells, *Prostate* **58**(3): 259–268.

Piccolo, S. (2008). p53 Regulation Orchestrates the TGF-β Response, *Cell* 133(5): 767–769.

Picus, J., Halabi, S., Rini, B., *et al.* (2003). The use of bevacizumab (B) with docetaxel (D) and estramustine (E) in hormone refractory prostate cancer (HRPC): initial results of CALGB 90006, *Proc. Am. Soc. Clin. Oncol.* **22**: 393.

Pienta, K. J., and Bradley. D. (2006). Mechanisms underlying the development of androgen-independent prostate cancer, *Clin. Cancer Res.* **12**(6): 1665–1671.

Pinkas, J., and Teicher, B. A. (2006). TGF-β in cancer and as a therapeutic target, *Biochem. Pharmacol.* **72**(5): 523–529.

Pisters, L. L., Pettaway, C. A., Troncoso, P., *et al.* (2004). Evidence that transfer of functional p53 protein results in increased apoptosis in prostate cancer, *Clin. Cancer Res.* **10**(8): 2587–2593.

Platz, E. A., Leitzmann, M. F., Visvanathan, K., *et al.* (2006). Statin drugs and risk of advanced prostate cancer, *J. Natl. Cancer Inst.* **98**: 1819–1825.

Platz, E. A., Rohrmann, S., Pearson, J. D., *et al.* (2005). Nonsteroidal anti-inflammatory drugs and risk of prostate cancer in the Baltimore Longitudinal Study of Aging, *Cancer Epidem. Biomar.* **14**: 390–396.

Plemper, R. K., and Wolf, D. H. (1999). Retrograde protein translocation: ERADication of secretory proteins in health and disease, *Trends Biochem. Sci.* **24**(7): 266–270.

Podar, K., Raje. N., and Anderson, K. C. (2007). Inhibition of the TGF-β Signaling Pathway in Tumor Cells. Targeted Interference with Signal Transduction Events, *Recent Res. Cancer* **172**: 77–97.

Pollak, M., Beamer, W., and Zhang, J. C. (1998). Insulin-like growth factors and prostate cancer, *Cancer Metast. Rev.* **17**(4): 383–390.

Pouyssegur, J., Dayan, F., and Mazure, N. M. (2006). Hypoxia signalling in cancer and approaches to enforce tumour regression, *Nature* **441**: 437–443.

Powell, M. J., Casimiro, M. C., Cordon-Cardo, C., *et al.* (2011). Disruption of a Sirt1-dependent autophagy checkpoint in the prostate results in prostatic intraepithelial neoplasia lesion formation, *Cancer Res.* **71**(3): 964–975.

Presta, L. G., Chen, H., O'Connor, S. J., *et al.* (1997). Humanization of an anti-vascular endothelial growth factor monoclonal antibody for the therapy of solid tumors and other disorders, *Cancer Res.* **57**: 4593–4599.

Price, N., and Dreicer, R. (2004). Phase I/II trial of bortezomib plus docetaxel in patients with advanced androgen-independent prostate cancer, *Clin. Prostate Cancer* **3**: 141–143.

Prins, G. S., Birch, L., and Greene, G. L. (1991). Androgen receptor localization in different cell types of the adult rat prostate, *Endocrinology* **129**(6): 3187.

Provenzano, P. P., Inman, D. R., Eliceri, K. W., *et al.* (2009). Matrix density-induced mechanoregulation of breast cell phenotype, signaling and gene expression through a FAK-ERK linkage, *Oncogene* **28**(49): 4326–4343.

Pruthi, R. S., Derksen, J. E., Moore, D., *et al.* (2006). Phase II trial of celecoxib in prostate-specific antigen recurrent prostate cancer after definitive radiation therapy or radical prostatectomy, *Clin. Cancer Res.* **12**(7:1): 2172–2177.

Pu, Y. S., Hour, T. C., Chuang, S. E., *et al.* (2004). Interleukin-6 is responsible for drug resistance and anti-apoptotic effects in prostatic cancer cells, *Prostate* **60**: 120–129.

Pylayeva, Y., Gillen, K. M., Gerald, W., *et al.* (2009). Ras- and PI3K-dependent breast tumorigenesis in mice and humans requires focal adhesion kinase signaling, *J. Clin. Invest.* **119**(2): 252–266.

Qi, W., Gao, S., and Wang, Z. (2008). Transcriptional regulation of the TGF-β1 promoter by androgen receptor, *Biochem. J.* **416**: 453–462.

Qiao, Y. L., Dawsey, S. M., Kamanger, F., *et al.* (2009). Total and cancer mortality after supplementation with vitamins and minerals: follow-up of the Linxian General Population Nutrition Intervention Trial, *J. Natl. Cancer Inst.* **101**: 507–518.

Qu, X., Zou, Z., Sun, Q., *et al.* (2007). Autophagy gene-dependent clearance of apoptotic cells during embryonic development, *Cell* **128**: 931–946.

Quayle, S. N., Mawji, N. R., Wang, J., *et al.* (2007). Androgen receptor decoy molecules block the growth of prostate cancer, *Proc. Natl. Acad. Sci. USA* **104**: 1331–1336.

Rabinovitz, I., Gipson, I. K., and Mercurio, A. M. (2001). Traction forces mediated by alpha6beta4 integrin: implications for basement membrane organization and tumor invasion, *Mol. Biol. Cell* **12**(12): 4030–4043.

Ramachandra, M., Atencio, I., Rahman, A., *et al.* (2002). Restoration of transforming growth factor Beta signaling by functional expression of smad4 induces anoikis, *Cancer Res.* **62**(21): 6045–6051.

Rao, R. V., Castro-Obregon, S., Frankowski, H., *et al.* (2002). Coupling endoplasmic reticulum stress to the cell death program. An Apaf-1-independent intrinsic pathway, *J. Biol. Chem.* **277**(24): 21836–21842.

Rao, R. V., Ellerby, H. M., and Bredesen, D. E. (2004). Coupling endoplasmic reticulum stress to the cell death program, *Cell Death Differ.* **11**(4): 372–380.

Rathkopf, D., Carducci, M. A., Morris, M. J., *et al.* (2008). Phase II trial of docetaxel with rapid androgen cycling for progressive noncastrate prostate cancer, *J. Clin. Oncol.* **26**: 2959–2965.

Ravid, D., Maor, S., Werner, H., *et al.* (2005). Caveolin-1 inhibits cell detachment-induced p53 activation and anoikis by upregulation of insulin-like growth factor-I receptors and signaling, *Oncogene* **24**: 1338–1347.

Razani, B., and Lisanti, M. P. (2001). Caveolin-deficient mice: insights into caveolar function human disease, *J. Clin. Invest.* **108**(11): 1553–1561.

Razani, B., Schlegel, A., Liu, J., *et al.* (2001). Caveolin-1, a putative tumour suppressor gene, *Biochem. Soc. Trans.* **29**(4): 494–499.

Reed, J. C. (1999). Mechanisms of apoptosis avoidance in cancer, *Curr. Opin. Oncol.* **11**: 68–75.

Reginato, M. J., Mills, K. R., Paulus, J. K., *et al.* (2003). Integrins and EGFR coordinately regulate the pro-apoptotic protein BIM to prevent anoikis, *Nat. Cell Biol.* **5**: 733–40.

Rehn, M., Hintikka, E., and Pihlajaniemi, T. (1994). Primary structure of the alpha 1 chain of mouse type XVIII collagen, partial structure of the corresponding gene, and comparison of the alpha 1(XVIII) chain with its homologue, the alpha 1(XV) collagen chain, *J. Biol. Chem.* **269**(19): 13929–13935.

Rennebeck, G., Martelli, M., and Kyprianou, N. (2005). Anoikis and survival connections in the reactive tumor microenvironment: significance in prostate cancer metastasis, *Cancer Res.* **63**: 11230–11235.

Reynolds, R. A., and Kyprianou, N. (2006). Growth factor signaling in prostatic growth: significance in tumour development and therapeutic targeting, *Br. J. Pharmacol.* 147: S144–S152.

Ricciardelli, C., Mayne, K., Sykes, P. J., *et al.* (1998). Elevated levels of versican but not decorin predict disease progression in early-stage prostate cancer, *Clin. Cancer Res.* 4(4): 963–71.

Richardson, P. G., Sonneveld, P., Schuster, M. W., *et al.* (2005). Bortezomib or high-dose dexamethasone for relapsed multiple myeloma, *N. Engl. J. Med.* 342: 2487–2498.

Rinaldo, F., Li, J., Wang, E., *et al.* (2007). RalA regulates vascular endothelial growth factor-C (VEGF-C) synthesis in prostate cancer cells during androgen ablation, *Oncogene* 26(12): 1731–1738.

Rittmaster, R. S., Fleshner, N., and Thompson, I. A. (2009). Pharmacological approaches to reducing the risk of prostate cancer, *Eur. Urol.* 55: 1064–1074.

Roberts, A. B., and Sporn, M. B. (1986). Transforming growth factor type beta: rapid induction of fibrosis and angiogenesis *in vivo* and stimulation of collagen formation *in vitro*, *Proc. Natl. Acad. Sci. USA* 83(12): 4167–4171.

Roberts, D. D. (1996). Regulation of tumor growth and metastasis by thrombospondin-1, *FASEB J.* 10(10): 1183–1191.

Roberts, J.T., and Essenhigh, D. M. (1986). Adenocarcinoma of prostate in 40-year old body-builder. *Lancet* 2(8509): 742.

Rodriguez-Manzaneque, J. C., Lane, T. F., Ortega, M. A., *et al.* (2001). Thrombospondin-1 suppresses spontaneous tumor growth and inhibits activation of matrix metallo-proteinase-9 and mobilization of vascular endothelial growth factor, *Proc. Natl. Acad. Sci. USA* 98(22): 12485–12490.

Rodriguez, J., and Lazebnik, Y. (1999). Caspase-9 and APAF-1 form an active holoenzyme, *Genes Dev.* 13(24): 3179–3184.

Rokhlin, O. W., Gudkov, A. V., Kwek, S., *et al.* (2000). p53 is involved in tumor necrosis factor-alpha-induced apoptosis in the human prostatic carcinoma cell line LNCaP, *Oncogene* 19(15): 1959–1968.

Romashkova, J. A., and Makarov, S. S. (1999). NF-kappaB is a target of AKT in anti-apoptotic PDGF signalling, *Nature* 401(6748): 86–90.

Rosini, P., Bonaccorsi, L., Baldi, E., *et al.* (2002). Androgen receptor expression induces FGF2, FGF-binding protein production, and FGF2 release in prostate carcinoma cells: role of FGF2 in growth, survival, and androgen receptor down-modulation, *Prostate* 53(4): 310–321.

Ross, J. S., Kallakury, B. V., Sheehan, C. E., *et al.* (2004). Expression of nuclear factor-kappa B and IκBα proteins in prostatic adenocarcinomas: correlation of nuclear factor-κB immunoreactivity with disease recurrence, *Clin. Cancer Res.* 10(7): 2466–2472.

Roth, D. M., Moseley, G. W., Glover, D., *et al.* (2007). A microtubule-facilitated nuclear import pathway for cancer regulatory proteins. *Traffic* 8: 673–686.

Rowley, D. R. (1998). What might a stromal response mean to prostate cancer progression? *Cancer Metast. Rev.* 17(4): 411–419.

Roy, N., Deveraux, Q. L., Takahashi, R., (1997). The c-IAP-1 and c-IAP-2 proteins are direct inhibitors of specific caspases, *EMBO J.* 16: 6914–6925.

Rubin, J. B., Kung, A. L., Klein, R. S., *et al.* (2003). A small-molecule antagonist of CXCR4 inhibits intracranial growth of primary brain tumors, *Proc. Natl. Acad. Sci. USA* **100**(23): 13513–13518.

Rudolfsson, S. H., and Bergh, A. (2009). Hypoxia drives prostate tumor progression and impairs the effectiveness of therapy, but can also promote cell death and serves as a therapeutic target, *Expert Opin. Ther. Tar.* **13**(2): 219–225.

Ruoslahti, E. (2002). Specialization of tumour vasculature, *Nat. Rev. Cancer* **2**: 83–90.

Russell, P. J., Bennett, S., and Stricken, P. (1998). Growth factor involvement in progression of prostate cancer, *Clin. Chem.* **44**(4): 705–723.

Rutkowski, D. T., and Kaufman, R. J. (2004). A trip to the ER: coping with stress, *Trends Cell Biol.* **14**(1): 20–28.

Sadar, M. D. (1999). Androgen-independent induction of prostate-specific antigen gene expression via cross-talk between the androgen receptor and protein kinase A signal transduction pathways, *J. Biol. Chem.* **274**: 7777–7783.

Sager, R., Sheng, S., Anisowicz, A., *et al.* (1994). RNA Genetics of Breast Cancer: Maspin as paradigm. *Cold Spring Harbor Symp. Quant. Biol.* **59**: 537–546.

Sager, R., Sheng, S., Pemberton, P., *et al.* (1996). Maspin: a tumor suppressing serpin, *Curr. Top Microbiol. Immunol.* **213**(1): 51–64.

Sakamoto, S., and Kyprianou, N. (2010). Targeting anoikis resistance in prostate cancer metastasis. *Mol. Aspects Med.* **31**(2): 205–214.

Sakamoto, S., McCann, R. O., Dhir, R., *et al.* (2010). Talin1 promotes tumor invasion and metastasis via focal adhesion signaling and anoikis resistance, *Cancer Res.* **70**(5): 1885–1895.

Sakamoto, S., Ryan, A. J., and Kyprianou, N. (2008). Targeting vasculature in urologic tumors: mechanistic and therapeutic significance, *J. Cell. Bioch.* **103**: 691–708.

Sakr, W. A., Grignon, D. J., Crissman, J. D., *et al.* (1994). High grade prostatic intraepithelial neoplasia (HGPIN) and prostatic adenocarcinoma between the ages of 20–69: an autopsy study of 249 cases. *In vivo*, **8**: 439–443.

Salm, S. N., Koikawa, Y., Ogilvie, V., *et al.* (2000). Generation of active TGF-beta by prostatic cell cocultures using novel basal and luminal prostatic epithelial cell lines, *J. Cell Physiol.* **184**(1): 70–79.

Sandford, N. L., Searle, J. W., and Kerr, J. F. (1984). Successive waves of apoptosis in the rat prostate after repeated withdrawal of testosterone stimulation, *Pathology* **16**(4): 406.

Sar, M., Lubahn, D. B., French, F. S., *et al.* (1990). Immunohistochemical localization of the androgen receptor in rat and human tissues, *Endocrinology* **127**(6): 3180–3186.

Sargeant, A. M., Rengel, R. C., Kulp, S. K., *et al.* (2008). OSU-HDAC42, a histone deacetylase inhibitor, blocks prostate tumor progression in the transgenic adenocarcinoma of the mouse prostate model, *Cancer Res.* **68**: 3999.

Saric, T., and Shain, S. A. (1998). Androgen regulation of prostate cancer cell FGF-1, FGF-2, and FGF-8: preferential down-regulation of FGF-2 transcripts, *Growth Factors* **16**(1): 69–87.

Satoh, T., Yang, G., Egawa, S., *et al.* (2003). Caveolin-1 expression is a predictor of recurrence-free survival in pT2N0 prostate carcinoma diagnosed in Japanese patients, *Cancer* **97**(5): 1225–1233.

Savagner, P. (2001). Leaving the neighborhood: molecular mechanisms involved during epithelial-mesenchymal transition, *Bioessays* **23**(10): 912–923.

Scaffidi, C., Schmitz, I., Krammer, P. H., *et al.* (1999). The role of c-FLIP in modulation of CD95-induced apoptosis, *J. Biol. Chem.* **274**: 1541–1548.

Scardino, P. T. (2003). The prevention of prostate cancer – the dilemma continues, *N. Engl. J. Med.*, **349**: 297–299.

Schafer-Hales, K., Iaconelli, J., Snyder, J. P., *et al.* (2007). Farnesyl transferase inhibitors impair chromosomal maintenance in cell lines and human tumors by compromising CENP-E and CENP-F function, *Mol. Cancer Ther.* **6**: 1317–1328.

Scheel-Toellner, D., Wang, K., Assi, L. K., *et al.* (2004). Clustering of death receptors in lipid rafts initiates neutrophil spontaneous apoptosis, *Biochem. Soc. Trans.* **32**(5): 679–681.

Scherr, D. S., Vaughan, E. D. Jr, Wei, J., *et al.* (1999). BCL-2 and p53 expression in clinically localized prostate cancer predicts response to external beam radiotherapy, *J. Urol.* **162**: 12–16.

Scheuner, D., Song, B., McEwen, E., *et al.* (2001). Translational control is required for the unfolded protein response and *in vivo* glucose homeostasis, *Molecular Cell* **7**: 1165–1176.

Schiemann, W. P., Blobe, G. C., Kalume, D. E., *et al.* (2002). Context-specific effects of fibulin-5 (DANCE/EVEC) on cell proliferation, motility, and invasion. Fibulin-5 is induced by transforming growth factor-beta and affects protein kinase cascades, *J. Biol. Chem.* **277**(30): 27367–27377.

Schimmer, A. D. (2004). Inhibitor of apoptosis proteins: translating basic knowledge into clinical practice, *Cancer Res.* **64**: 7183–7190.

Schlessinger, K., and Hall, A. (2004). GSK-3beta sets Snail's pace, *Nat. Cell Biol.* **6**(10): 913–915.

Schlingensiepen. K. H., Bicshof, A., Egger, T., *et al.* (2004). The TGF-beta1 antisense oligonucleotide AP 11014 for the treatment of non-small cell lung, colorectal and prostate cancer: Preclinical studies, *J. Clin. Oncol.* **22**(14s): 3132.

Schor, S. L., Schor, A. M., and Rushton, G. (1988). Fibroblasts from cancer patients display a mixture of both foetal and adult-like phenotypic characteristics, *J. Cell Sci.* **90**(3): 401–407.

Schroder, M., and Kaufman, R. J. (2005a). The mammalian unfolded protein response, *Annu. Rev. Biochem.* **74**: 739–789.

Schroder, M., and Kaufman, R. J. (2005b). ER stress and the unfolded protein response, *Mutat. Res.* **569**(1–2): 29–63.

Schwarze, S. R., Lin, E. W., Christian, P. A., *et al.* (2008). Intracellular death platform steps in: targeting prostate tumors via endoplasmic reticulum (ER) apoptosis, *Prostate* **68**: 1615–1623.

Scorrano, L., Oakes, S. A., Opferman, J. T., *et al.* (2003). BAX and BAK regulation of endoplasmic reticulum Ca2+: a control point for apoptosis, *Science* **300**(5616): 135–139.

Scott, W. W. (1953). Endocrine management of disseminated prostatic cancer, including bilateral adrenalectomy and hypophysectomy, *Trans. Am. Ass. Genitourinary Surg.* **44**: 101–104.

Seemayer, T. A., Lagace, R., Schurch, W., *et al.* (1979). Myofibroblasts in the stroma of invasive and metastatic carcinoma: a possible host response to neoplasia, *Am. J. Surg. Pathol.* **3(6)**: 525–533.

Sells, S. F., Wood, D. P. Jr, Joshi-Barve, S. S., *et al.* (1994). Commonality of the gene programs induced by effectors of apoptosis in androgen-dependent and -independent prostate cells, *Cell Growth Differ.* **5**: 457–466.

Senft, J., Helfer, B., and Frisch, S, M. (2007). Caspase-8 Interacts with the p85 Subunit of Phosphatidylinositol 3-Kinase to Regulate Cell Adhesion and Motility, *Cancer Res.* **67**: 11505.

Seol, D. W., Li, J., Seol, M-H., *et al.* (2001). Signaling events triggered by tumor necrosis factor-related apoptosis-inducing ligand (TRAIL): Caspase-8 is required for TRAIL-induced apoptosis, *Cancer Res.* **51**: 1138.

Serini, G., Valdembri, D., Zanivan, S., *et al.* (2003). Class 3 semaphorins control vascular morphogenesis by inhibiting integrin function, *Nature* **424(24)**: 391–397.

Sfanos, K. S., Wilson, B. A., De Marzo, A. M., *et al.* (2009). Acute inflammatory proteins constitute the organic matrix of prostatic corpora amylacea and calculi in men with prostate cancer, *Proc. Natl. Acad. USA* **106**: 3443–3448.

Shabsigh, A., Chang, D. T., Heitjan, D. F., *et al.* (1998). Rapid reduction in blood flow to the rat ventral prostate gland after castration: preliminary evidence that androgens influence prostate size by regulating blood flow to the prostate gland and prostatic endothelial cell survival, *Prostate* **36(3)**: 201–206.

Shah, A. H., Tabayoyong, W. B., Kundu, S. D., *et al.* (2002). Suppression of tumor metastasis by blockade of transforming growth factor beta signaling in bone marrow cells through a retroviral-mediated gene therapy in mice, *Cancer Res.* **62(24)**: 7135–7138.

Shah, R. B., Mehra, R., Chinnaiyan, A. M., *et al.* (2004). Androgen-independent prostate cancer is a heterogeneous group of diseases: lessons from a rapid autopsy program, *Cancer Res.* **64**: 9209–9216.

Shahinian, V. B., Kuo, Y. F., Freeman, J. L., *et al.* (2005). Risk of fracture after androgen deprivation for prostate cancer, *N. Engl. J. Med.* **352**: 154–164.

Shang, Y., Myers, M., and Brown, M. (2002). Formation of the androgen receptor transcription complex, *Mol. Cell* **9(3)**: 601–610.

Shariat, S. F., Andrews, B., Kattan, M. W., *et al.* (2001). Plasma levels of interleukin-6 and its soluble receptor are associated with prostate cancer progression and metastasis, *Urology* **58**: 1008–1015.

Shariat, S. F., Kattan, M. W., Traxel, E., *et al.* (2004a). Association of pre- and postoperative plasma levels of transforming growth factor β (1) and interleukin 6 and its soluble receptor with prostate cancer progression, *Clin. Cancer Res.* **10(6)**: 1992–1999.

Shariat, S. F., Menesses-Diaz, A., Kim, I. Y., *et al.* (2004b). Tissue expression of transforming growth factor-beta1 and its receptors: correlation with pathologic features and biochemical progression in patients undergoing radical prostatectomy, *Urology* **63(6)**: 1191–1197.

Sharifi, N., Hamel, E., Lill, M. A., *et al.* (2007). A bifunctional colchicinoid that binds to the androgen receptor, *Mol. Cancer Ther.* **6**: 2328–2336.

Shaw, Y. J., Yang, Y. T., Garrison, J. B., *et al.* (2004). Pharmacological exploitation of the α1-adrenoreceptor antagonist doxazosin to develop a novel class of antitumor agents that block intracellular protein kinase B/Akt activation, *J. Med. Chem.* **47**(18): 4453–4462.

Shen, J., Chen, X., Hendershot, L., *et al.* (2002). ER stress regulation of ATF6 localization by dissociation of BiP/GRP78 binding and unmasking of Golgi localization signals, *Dev. Cell* **3**(1): 99–111.

Shi, H. Y., Liang, R., Templeton, N. S., *et al.* (2002). Inhibition of breast tumor progression by systemic delivery of the maspin gene in a syngeneic tumor model, *Mol. Ther.* **5**(6): 7555–7561.

Shi, Y., and Massague, J. (2003). Mechanisms of TGF-β signaling from cell membrane to the nucleus, *Cell* **113**: 685–700.

Shimura, S., Yang, G., Ebara, S., *et al.* (2000). Reduced infiltration of tumor-associated macrophages in human prostate cancer: association with cancer progression, *Cancer Res.* **60**: 5857–5861.

Shuch, B., Mikhail, M., Satagopan, J., *et al.* (2004). Racial disparity of epidermal growth factor receptor expression in prostate cancer, *J. Clin. Oncol.* **22**(23): 4725–4729.

Shukla, S., and Gupta, S. (2004). Suppression of constitutive and tumor necrosis factor alpha-induced nuclear factor (NF)-κB activation and induction of apoptosis by apigenin in human prostate carcinoma PC-3 cells: correlation with down-regulation of NF-κB-responsive genes, *Clin. Cancer Res.* **10**(9): 3169–3178.

Siegel, P. M., and Massague, J. (2003). Cytostatic and apoptotic actions of TGF-beta in homeostasis and cancer, *Nat. Rev. Cancer* **3**(11): 807–821.

Siiteri, P. K., and Wilson, J. D. (1974). Testosterone formation and metabolism during male sexual differentiation in the human embryo, *J. Clin. Endocrinol. Metab.* **38**(1): 1113–1125.

Sikes, R. A., Nicholson, B. E., Koeneman, K. S., *et al.* (2004). Cellular inter-actions in the tropism of prostate cancer to bone, *Int. J. Cancer* **110**(4): 497–503.

Simental, J. A., Sar, M., Lane, M. V., *et al.* (1991). Transcriptional activation and nuclear targeting signals of the human androgen receptor, *J. Biol. Chem.* **266**: 510–518.

Simon, K., and Toomre, D. (2000). Lipid rafts and signal transduction, *Nat. Rev. Mol. Cell Biol.* **1**(1): 31–39.

Simpson, C. D., Anyiwe, K., and Schimmer, A. D. (2008). Anoikis resistance and tumor metastasis. *Cancer Lett.* **272**: 177–185.

Sipkins, D. A., Cheresh, D. A., Kazemi, M. R., *et al.* (1998). Detection of tumor angiogenesis *in vivo* by alphaVbeta3-targeted magnetic resonance imaging, *Nat. Med.* **4**: 623–626.

Skinner, H. D., Zheng, J. Z., Fang, J., *et al.* (2004). Vascular endothelial growth factor transcriptional activation is mediated by hypoxia-inducible factor 1α, HDM2, and p70S6K1 in response to phosphatidylinositol 3-kinase/AKT signaling, *J. Biol. Chem.* **279**(44): 45643–45651.

Sklar, G. N., Eddy, H. A., Jacobs, S. C., *et al.* (1993). Combined antitumor effect of suramin plus irradiation in human prostate cancer cells: the role of apoptosis, *J. Urol.* **150**: 1526–1532.

Small, E. J., Halabi, S., Ratain, M. J., *et al.* (2002). Randomized study of three different doses of suramin administered with a fixed dosing schedule in patients with advanced prostate cancer: results of Intergroup Cancer and Leukemia Group B 9480, *J. Clin. Oncol.* **20**: 3369–3375.

Small, E. J., Meyer, M., Marshall, M. E., *et al.* (2000). Suramin therapy for patients with symptomatic hormone-refractory prostate cancer: results of a randomized phase III trial comparing suramin plus hydrocortisone to placebo plus hydrocortisone, *J. Clin. Oncol.* **18**: 1440–1450.

Small, E. J., Reese, D. M., Um, B., *et al.* (1999). Therapy of advanced prostate cancer with granulocyte macrophage colony-stimulating factor1, *Clin. Cancer Res.* **5**: 1738.

Smith, M. C. P., Luker, K. E., Garbow, J. R., *et al.* (2004). CXCR4 regulates growth of both primary and metastatic breast cancer, *Cancer Res.* **64**: 8604.

Smith, M. R., Manola, J., Kaufman, D. S., *et al.* (2006). Celecoxib versus placebo for men with prostate cancer and a rising serum prostate-specific antigen after radical prostatectomy and/or radiation therapy, *J. Clin. Oncol.* **24**: 2723–2728.

Smith, P. C., and Keller, E. T. (2001). Anti-interleukin-6 monoclonal antibody induces regression of human prostate cancer xenografts in nude mice, *Prostate* **48**: 47–53.

Song, K., Cornelius, S. C., Reiss, M., *et al.* (2003). Insulin-like growth factor-I inhibits transcriptional responses of transforming growth factor-beta by share 3-kinase/Akt-dependent suppression of the activation of Smad3 but not Smad2, *J. Biol. Chem.* **278**(40): 38342–38351.

Sonmez, H., Suer, S., Karaarslan, I., *et al.* (1995). Tissue fibronectin levels of human prostatic cancer, as a tumor marker, *Cancer Biochem. Biophys.* **15**(2): 107–110.

Sooriakumaran, P., Langley, S. E., Laing, R. W., *et al.* (2007). COX-2 inhibition: a possible role in the management of prostate cancer? *J. Chemother.* **19**: 21–32.

Sorrentino, A., Thakur, N., Grimsby, S., *et al.* (2008). The type I TGF- β receptor engages TRAF6 to activate TAK1 in a receptor kinase-independent manner, *Nat. Cell Biol.* **10**: 1199–1207.

Sosic, D., Richardson, J. A., Yu, K., *et al.* (2003). Twist regulates cytokine gene expression through a negative feedback loop that represses NF-kappaB activity, *Cell* **112**(2): 169–180.

Soussi, T., Dehouche, K., and Beroud, C. (2000). p53 website and analysis of p53 gene mutations in human cancer: forging a link between epidemiology and carcinogenesis, *Hum. Mutat.* **15**: 105–113.

Sporn, M. B., and Roberts, A. B. (1992). Transforming growth factor-beta: recent progress and new challenges, *J. Cell Biol.* **119**(5): 1017–1021.

Sprick, M., Weigand, M. A., Rieser, E., *et al.* (2000). FADD/MORT1 and caspase-8 are recruited to TRAIL receptors 1 and 2 and are essential for apoptosis mediated by TRAIL receptor 2, *Immunity* **12**: 599–609.

Srinivasula, S. M., Ahmad, M., Fernandes-Alnemri, T., *et al.* (1996). Molecular ordering of the Fas-apoptotic pathway: the Fas/APO-1 protease Mch5 is a CrmA-inhibitable protease that activates multiple Ced-3/ICE-like cysteine proteases, *Proc. Natl. Acad. Sci. USA* **93**: 14486–14491.

Srivastava, R. K., Srivastava, A. R., Korsmeyer, S. J., *et al.* (1998). Involvement of microtubules in the regulation of Bcl2 phosphorylation and apoptosis through cyclic AMP-dependent protein kinase, *Mol. Cell Biol.* **18**(6): 3509–3517.

Stadler, W. M., Cao, D., Vogelzang, N. J., *et al.* (2004). A randomized Phase II trial of the antiangiogenic agent SU5416 in hormone-refractory prostate cancer, *Clin. Cancer Res.* **10**: 3365–3370.

Stanbrough, M., Bubley, G. J., Ross, K. *et al.* (2006). Increased expression of genes converting adrenal androgens to testosterone and androgens in androgen-independent prostate cancer, *Cancer Res.* **66**: 2815–2825.

Stanbrough, M., Leav, I., Kwan, P. W. L., *et al.* (2001). Prostatic intraepithelial neoplasia in mice expressing an androgen receptor transgene in prostate epithelium, *Proc. Nat. Acad. Sci.* **98(19)**: 10823–10828.

Stattin, P., Damber, J. E., Modig, H., *et al.* (1996). Pretreatment p53 immunoreactivity does not infer radioresistance in prostate cancer patients, *Int. J. Radiat. Oncol. Biol. Phys.* **35**: 885–889.

Stattin, P., Rinaldi, S., Biessy, C., *et al.* (2004). High levels of circulating insulin-like growth factor-I increase prostate cancer risk: a prospective study in a population-based nonscreened cohort, *J. Clin. Oncol.* **22(15)**: 3104–3112.

Stellmach, V., Volpert, O. V., Crawford, S. E., *et al.* (1996). Tumour suppressor genes and angiogenesis: the role of TP53 in fibroblasts, *Eur. J. Cancer* **32A(14)**: 2394–2400.

Stern, D. F. (2004). More than a marker: phosphorylated Akt in prostatic carcinoma, *Clin. Cancer Res.* **10**: 6407–6410.

Streit, M., Velasco, P., Brown, L. F., *et al.* (1999). Overexpression of thrombospondin-1 decreases angiogenesis and inhibits the growth of human cutaneous squamous cell carcinomas, *Am. J. Pathol.* **155(2)**: 441–452.

Sugahara, K. N., Teesalu, T., Karmali, P. P., *et al.* (2009). Tissue-penetrating delivery of compounds and nanoparticles into tumors, *Cancer Cell* **16(6)**: 510–520.

Suh, J., and Rabson, A. B. (2004). NF-κB activation in human prostate cancer: important mediator or epiphenomenon? *J. Cell Biochem.* **91**: 100–117.

Sumitomo, M., Tachibana, M., Nakashima, J., *et al.* (1999). An essential role for nuclear factor kappa B in preventing TNF-alpha-induced cell death in prostate cancer cells, *J. Urol.* **161(2)**: 674–679.

Sun, S. Y., Hail, N. Jr, and Lotan, R. (2004). Apoptosis as a novel target for cancer chemoprevention, *J. Natl. Cancer Inst.* **96(9)**: 662–672.

Sun, Y. X., Wang, J., Shelburne, C. E., *et al.* (2003). Expression of CXCR4 and CXCL12 (SDF-1) in human prostate cancers (PCa) *in vivo*, *J. Cell Biochem.* **89(3)**: 462–473.

Sung, S-Y., and Chung, L. W. K. (2002). Prostate tumor-stroma interaction: molecular mechanisms and opportunities for therapeutic targeting, *Differentiation* **70(9-10)**: 506–521.

Sunters, A., Fernandez de Mattos, S., Stahl, M., *et al.* (2003). FoxO3a transcriptional regulation of BIM controls apoptosis in Paclitaxel-treated breast cancer cell lines, *J. Biol. Chem.* **278**: 49795–49805.

Suzuki, H., Okihara, H., Fujisawa, M., *et al.* (2008). Alternative nonsteroidal antiandrogen therapy for advanced prostate cancer that relapsed after initial maximum androgen blockade, *J. Urology* **180(3)**: 921–927.

Sweeney, C., Liu, G., Yiannoutsos, C., *et al.* (2005). A phase II multicenter, randomized, double-blind, safety trial assessing the pharmacokinetics, pharmacodynamics, and efficacy of oral 2-methoxyestradiol capsules in hormone-refractory prostate cancer, *Clin. Cancer Res.* **11**: 6625.

Szostak, M., Kaur, P., Amin, P., *et al.* (2001). Apoptosis and BCL-2 expression in prostate cancer: significance in clinical outcome after brachytherapy, *J. Urology*, **165**: 2126–2130.

Szostak, M., and Kyprianou, N. (2000). Radiation-induced apoptosis: predictive and therapeutic significance in radiotherapy of prostate cancer, *Oncol. Rep.* **7(4)**: 699–706.

Tahir, S. A., Yang, G., Ebara, S., *et al.* (2001). Secreted caveolin-1 stimulates cell survival/ clonal growth and contributes to metastasis in androgen-insensitive prostate cancer, *Cancer Res.* **61(10)**: 3882–3885.

Tahmatzopoulos, A., and Kyprianou, N. (2004). Apoptotic impact of $\alpha_1$-adrenoceptor antagonists on prostate cancer growth: a myth or an inviting reality? *Prostate* **59**: 91–100.

Tahmatzopoulos, A., Sheng, S., and Kyprianou, N. (2005). Maspin sensitizes prostate cancer cells to doxazosin-induced apoptosis, *Oncogene* **24**: 5375–5383.

Tait, W. G., and Green, D. R. (2010). Mitochondria and cell death: outer membrane permeabilization and beyond. *Nat. Rev. Mol. Cell Bio.* **11**: 621–632.

Takeichi, M. (1991). Cadherin cell adhesion receptors as a morphogenetic regulator, *Science* **251**: 1451–1455.

Tanaka, M., Suda, T., Haze, K., *et al.* (1996). Fas ligand in human serum, *Nat. Med.* **2**: 317–322.

Tang, D. G., and Porter, A. T. (1997). Target to apoptosis: a hopeful weapon for prostate cancer, *Prostate* **32(4)**: 284.

Tannock, I. F., de Wit, R., Berry, W. R., *et al.* (2004). Docetaxel plus prednisone or mitoxantrone plus prednisone for advanced prostate cancer, *N. Engl. J. Med.* **351**: 1502–1512.

Tao, K. S., Dou, K. F., and Wu, X. A. (2004). Expression of angiostatin cDNA in human hepatocellular carcinoma cell line SMMC-7721 and its effect on implanted carcinoma in nude mice, *World J. Gastroenterol.* **10(10)**: 1421–1424.

Taplin, M. E. (2007). Drug insight: role of the androgen receptor in the development and progression of prostate cancer, *Nat. Clin. Pract. Oncol.* **4**: 236–244.

Taplin, M. E., Bubley, G. J., Ko, Y. J., *et al.* (1999). Selection for androgen receptor mutations in prostate cancers treated with androgen antagonist, *Cancer Res.* **59**: 2511–2515.

Tarin, D., Thompson, E. W., and Newgreen, D. F. (2005). The fallacy of epithelial mesenchymal transition in neoplasia, *Cancer Res.* **65(14)**: 5996–6001.

Taylor, R. C., Cullen, S.P., and Martin, S. J. (2008). Apoptosis: controlled demolition at the cellular level. *Nat. Rev. Mol. Cell Biol.* **9**: 231–241.

Teicher, B. A. (2001). Malignant cells, directors of the malignant process: role of transforming growth factor-β, *Cancer Metast. Rev.* **20(1–2)**: 133–143.

Thakkar, S. G., Choueiri, T. K., and Garcia, J. A. (2006). Endothelin receptor antagonists: rationale, clinical development and role in prostate cancer therapeutics, *Curr. Oncol. Rep.* **8(2)**: 108–113.

The Alpha-Tocopherol Beta Carotene Cancer Prevention Study Group (1994). The effect of vitamin E and beta carotene on the incidence of lung cancer and other cancers in male smokers, *N. Engl. J. Med.* **330**: 1029–1035.

Thiery, J. P. (2002). Epithelial-mesenchymal transitions in tumour progression, *Nat. Rev. Cancer* **2(6)**: 442–454.

Thiery, J. P. (2003). Epithelial-mesenchymal transitions in development and pathologies, *Curr. Opin. Cell Biol.* **15(6)**: 740–746.

Thiery, J. P., and Sleeman, J. P. (2006). Complex networks orchestrate epithelial–mesenchymal transitions, *Nat. Rev. Mol. Cell Bio.* **7**: 131–142.

Thomas, D. A., and Massague, J. (2005). TGF-β directly targets cytotoxic T cell functions during tumor evasion of immune surveillance, *Cancer Cell* **8**: 369–380.

Thomas, G., Jacobs, K. B., Yeager, M., *et al.* (2008). Multiple loci identified in a genome-wide association study of prostate cancer, *Nat. Genet.* **40**: 310–315.

Thomas, G. V., Horvath, S., Smith B. L., *et al.* (2004). Antibody-based profiling of the phosphoinositide 3-kinase pathway in clinical prostate cancer, *Clin. Cancer Res.* **10**: 8351–8356.

Thome, M., Schneider, P., Hofmann, K., *et al.* (1997). Viral FLICE-inhibitory proteins (FLIPs) prevent apoptosis induced by death receptors, *Nature* **386**: 517–521.

Thompson, E. W., Newgreen, D. F., and Tarin, D. (2005). Carcinoma invasion and metastasis: a role for epithelial-mesenchymal transition? *Cancer Res.* **65(14)**: 5991–5995.

Thompson, I. M., Ankerst, D. P., Chi, C., *et al.* (2006a). Assessing prostate cancer risk: results from the Prostate Cancer Prevention Trial, *J. Natl. Cancer Inst.* **98(8)**: 529–534.

Thompson, I. M., Chi, C., Ankerst, D. P., *et al.* (2006b). Effect of finasteride on the sensitivity of PSA for detecting prostate cancer, *J. Natl. Cancer Inst.* **98(16)**: 1128–1133.

Thompson, I. M., Goodman, P. J., Tangen, C. M., *et al.* (2003). The influence on the development of prostate cancer, *N. Engl. J. Med.* **349(3)**: 215–224.

Thompson, I. M., Pauler, D. K., Ankerst. D. P., *et al.* (2007). Prediction of prostate cancer for patients receiving finasteride: results from the Prostate Cancer Prevention Trial, *J. Clin. Oncol.* **25(21)**: 3076–3081.

Thompson, I. M., Pauler, D. K., Goodman, P. J., *et al.* (2004). Prevalence of prostate cancer among men with a prostate-specific antigen level < or = 4.0 ng per milliliter, *N. Engl. J. Med.* **350(22)**: 2239–2246.

Thorpe, J. A., Christian, P. A., and Schwarze, S. R. (2008). Proteasome inhibition blocks caspase-8 degradation and sensitizes prostate cancer cells to death receptor-mediated apoptosis, *Prostate* **68**: 200–209.

Thuault, S., Tan, E-J., Peinado, H., *et al.* (2008). HMGA2 and Smads co-regulate SNAIL1 expression during induction of epithelial-to-mesenchymal transition, *J. Biol. Chem.* **283**: 33437–33446.

Timpl, R., Sasaki, T., Kostka, G., *et al.* (2003). Fibulins: a versatile family of extracellular matrix proteins, *Nat. Rev. Mol. Cell Biol.* **4(6)**: 479–489.

Titus, M. A., Schell, M. J., Lih, F. B., *et al.* (2005). Testosterone and dihydrotestosterone tissue levels in recurrent prostate cancer, *Clin. Cancer Res.* **11**: 4653.

Tlsty, T. D., and Hein, P. W. (2001). Know thy neighbor: stromal cells can contribute oncogenic signals, *Curr. Opin. Genet. Dev.* **11(1)**: 54–59.

Tombal, B., Denmeade, S. R., Gillis, J. M., *et al.* (2002). A supramicromolar elevation of intracellular free calcium ($[Ca^{2+}]i$) is consistently required to induce the execution phase of apoptosis, *Cell Death Differ.* **9(5)**: 561–573.

Tomita, K., van Bokhoven, A., van Leenders, G. J., *et al.* (2000). Cadherin switching in human prostate cancer progression, *Cancer Res.* **60**(13): 3650–3654.

Tomita, Y., Marchenko, N., Erster, A., *et al.* (2006). WT p53, but not tumor-derived mutants, bind to BCL-2 via the DNA binding domain and induce mitochondrial permeabilization, *J. Biol. Chem.* **281**: 8600–8606.

Torring, N., Dagnaes-Hansen, F., Sorenson, B. S., *et al.* (2003). ErbB1 and prostate cancer: ErbB1 activity is essential for androgen-induced proliferation and protection from the apoptotic effects of LY294002, *Prostate* **56**(2): 142–149.

Tran, C., Ouk, S., Clegg, N. J., *et al.* (2009). Development of a second–generation antiandrogen for treatment of advanced prostate cancer, *Science* **324**: 787–790.

Treiman, M., Caspersen, C., and Christensen, S. B. (1998). A tool coming of age: thapsigargin as an inhibitor of sarco-endoplasmic reticulum $Ca(2+)$-ATPases, *Trends Pharmacol. Sci.* **19**(4): 131–135.

Tsai, B., Ye, Y., and Rapoport, T. A. (2002). Retro-translocation of proteins from the endoplasmic reticulum into the cytosol, *Nat. Rev. Mol. Cell Biol.* **3**(4): 246–255.

Tso C. L., McBride W. H., Sun, J., *et al.* (2000). Androgen deprivation induces selective outgrowth of aggressive hormone refractory prostate cancer clones expressing distinct cellular and molecular properties not present in parental androgen-dependent cancer cells. *Cancer J.* **6**: 213–214.

Tsuji, T., Ibaragi, S., and Hu, G-F. (2009). Epithelial-mesenchymal transition and cell cooperativity in metastasis, *Cancer Res.* **69**: 7135–7139.

Tsujimoto, Y. (1989). Stress-resistance conferred by high level of bcl-2alpha protein in human B lypmphoblastoid cell, *Oncogene* **4**: 1331–1336.

Tsujimoto, Y., Cossman, J., Jaffe, E., *et al.* (1985). Involvement of the bcl-2 gene in human follicular lymphoma, *Science* **228**: 1440–1443.

Tuxhorn, J. A., Ayala, G. E., and Rowley, D. R. (2001). Reactive stroma in prostate cancer progression, *J. Urol.* **166**(6): 2472–2483.

Tuxhorn, J. A., Ayala, G. E., Smith, M. J., *et al.* (2002a). Reactive stroma in human prostate cancer: induction of myofibroblast phenotype and extracellular matrix remodeling, *Clin. Cancer Res.* **8**(9): 2912–2923.

Tuxhorn, J. A., McAlhany, S. J., Dang, T. D., *et al.* (2002b). Stromal cells promote angiogenesis and growth of human prostate tumors in a differential reactive stroma (DRS) xenograft model, *Cancer Res.* **62**(11): 3298–3307.

Tuxhorn, J. A., McAlhany, S. J., Yang, F., *et al.* (2002c). Inhibition of transforming growth factor-beta activity decreases angiogenesis in a human prostate cancer-reactive stroma xenograft model, *Cancer Res.* **62**(21): 6021–6025.

Ueda, T., Bruchovsky, N., and Sadar, M. D. (2002). Activation of the N-terminus of the androgen receptor by interleukin-6 via MAPK and STAT3 signal transduction pathways cells, *J. Biol. Chem.* **277**: 7076–7085.

Ueda, T., Mawji, N. R., Bruchovsky, N., *et al.* (2002). Ligand-independent activation of the androgen receptor by interleukin-6 and the role of steroid receptor coactivator-1 in prostate cancer cells, *J. Biol. Chem.* **277**(41): 38087–38094.

Umansky, S.R. (1982). The genetic program of cell death. Hypothesis and some applications: transformation, carcinogenesis, ageing, *J. Theor. Biol.* **97**(4): 591–602.

Umansky, S. R., Korol, B. A., and Lelipovich, P. A. (1981). *In vivo* DNA degradation in thymocytes of γ-irradiated or hydrocortisone-treated rats, *Biochim. Biophys. Acta.* 655: 9–17.

Urquidi, V., Sloan, D., Kawai, K., *et al.* (2002). Contrasting expression of thrombospondin-1 and osteopontin correlates with absence or presence of metastatic phenotype in an isogenic model of spontaneous human breast cancer metastasis, *Clin. Cancer Res.* 8(1): 61–74.

Valentijn, A. J., Zouq, N., and Gilmore, A. P. (2004). Anoikis, *Biochem. Soc. Trans.* 32(3): 421–425.

Valmiki, M. G., and Ramos, J. W. (2009). Death effector domain-containing proteins, *Cell Mol. Life Sci.* 66: 814–830.

Van der Poel, H. G. (2005). Androgen receptor and TGFbeta1/Smad signaling are mutually inhibitory in prostate cancer, *Eur. Urol.* 48(6): 1051–1058.

Vardouli, L., Moustakas, A., and Stournaras, C. (2005). LIM-kinase 2 and cofilin phosphorylation mediate actin cytoskeleton reorganization induced by transforming growth factor-β, *J. Biol. Chem.* 280: 11448–11457.

Vaupel, P., and Mayer, A. (2007). Hypoxia in cancer significance and impact on clinical outcome, *Cancer Metast. Rev.* 26(2): 225–239.

Vaux, D., Cory, S., and Adams, J. M. (1998). Bcl-2 gene promotes haematopoietic cell survival and cooperates with c-myc to immortalize pre-B cells, *Nature* 335: 440–442.

Vega, S., Morales, A. V., Ocana, O. H., *et al.* (2004). Snail blocks the cell cycle and confers resistance to cell death, *Genes Dev.* 18(10): 1131–1143.

Veis, D., Sorenson, C. M., Shutter, J. R., *et al.* (1993). BCL-2-deficient mice demonstrate fulminant lymphoid apoptosis, polycystic kidneys, and hypopigmented hair, *Cell* 75: 229–240.

Veitonmaki, N., Cao, R., Wu, L. H., *et al.* (2004). Endothelial cell surface ATP synthase-triggered caspase-apoptotic pathway is essential for k1-5-induced anti-angiogenesis, *Cancer Res.* 64(10): 3679–3686.

Verhagen, A., Ekert, P. G., Pakusch, M., *et al.* (2000). Identification of DIABLO, a mammalian protein that promotes apoptosis by binding to an antagonizing IAP protein, *Cell* 102: 43–53.

Viatour, P., Merville, M. P., Bours, V., *et al.* (2005). Phosphorylation of NF-κB and IκB proteins: implications in cancer and inflammation, *Trends Biochem. Sci.* 30(1): 43–52.

Vicentini, C., Festuccia, C., Gravina, G. L., *et al.* (2003). Prostate cancer cell proliferation is strongly reduced by the epidermal growth factor receptor tyrosine kinase inhibitor ZD1839 *in vitro* on human cell lines and primary cultures, *J. Cancer Res. Clin. Oncol.* 129(3): 165–174.

Vishnu, P., and Tan, W. W. (2010). Update on options for treatment of metastatic castration-resistant prostate cancer, *Onco. Targets Ther.* 3: 39–51.

Volpert, O. V., Zaichuk, T., Zhou, W., *et al.* (2002). Inducer-stimulated Fas targets activated endothelium for destruction by anti-angiogenic thrombospondin-1 and pigment epithelium-derived factor, *Nat. Med.* 8(4): 349–357.

Wakoshi, N. N., Lee, A-H., Vallabhajosyula, P., *et al.* (2003). Plasma cell differentiation and the unfolded protein response intersect at the transcription factor XBP-1, *Nat. Immunol.* 4: 321–329.

Walden, P. D., Globina, Y., and Nieder, A. (2004). Induction of anoikis by doxazosin in prostate cancer cells is associated with activation of caspase-3 and a reduction of focal adhesion kinase, *Urol. Res.* 32(4): 261–265.

Wang, G., Wang. J., and Sadar, M. D. (2008). Expand+crosstalk between the androgen receptor and β-catenin in castrate-resistant prostate cancer, *Cancer Res.* 68: 9918.

Wang, H., Song, K., Sponseller, T. L., *et al.* (2005). Novel function of androgen receptor-associated protein 55/Hic-5 as a negative regulator of Smad3 signaling, *J. Biol. Chem.* 280(7): 5154–5162.

Wang, J., Chun, H. J., Wong, W., *et al.* (2001). Caspase-10 is an initiator caspase in death receptor signaling, *Proc. Natl. Acad. Sci. USA* 98(24): 13884–13888.

Wang, J-D., Takahara, S., Nonomura, N., *et al.* (1999). Early induction of apoptosis in androgen – independent prostate cancer cell line by FTY720 requires caspase-3 activation, *Prostate* 40(1): 50–55.

Wang, L., Zuercher, W. J., Consier, T. G., *et al.* (2006). X-ray crystal structures of the estrogen-related receptor-gamma ligand binding domain in three functional states reveal the molecular basis of small molecule regulation, *J. Biol. Chem.* 281: 37773–37781.

Wang, M. H., Helzlsouer, K. J., Smith, M. W., *et al.* (2009). Association of IL10 and other immune response- and obesity-related genes with prostate cancer in CLUE II, *Prostate* 69: 874–885.

Wang, P., Gilmore, A. P., and Streuli, C. H. (2004). BIM is an apoptosis sensor that responds to loss of survival signals delivered by epidermal growth factor but not those provided by integrins, *J. Biol. Chem.* 279: 41280–41285.

Wang, S., Gao, J., Lei, Q., *et al.* (2003). Prostate-specific deletion of the murine Pten tumor suppressor gene leads to metastatic prostate cancer, *Cancer Cell* 4(3): 209–221.

Wang, X., Yin, L., Rao, P., *et al.* (2007). Targeted treatment of prostate cancer. *J. Cell Biochem.* 102: 571–579.

Wang, X. Z., Lawson, B., Brewer, J. W., *et al.* (1996). Signals from the stressed endoplasmic reticulum induce C/EBP-homologous protein (CHOP/GADD153), *Mol. Cell. Biol.* 16(8): 4273–4280.

Wang, Y., Shibasaki, F., and Mizuno, K. (2005). Calcium signal-induced cofilin dephosphorylation is mediated by Slingshot via calcineurin, *J. Biol. Chem.* 280: 12683–12689.

Waters, D. J., Shen, S., Cooley, D. M., *et al.* (2003). Effects of dietary selenium supplementation on DNA damage and apoptosis in canine prostate, *J. Natl. Cancer Inst.* 95: 237–241.

Weeraratna, A. T., Arnold, J. T., George, D. J., *et al.* (2000). Rational basis for Trk inhibition therapy for prostate cancer, *Prostate* 45(2): 140–148.

Wei, M., Zong, W. Z., Cheng, E. H., *et al.* (2001). Proapoptotic BAX and BAK: a requisite gateway to mitochondrial dysfunction and death, *Science* 292: 727–730.

Weichert, W., Röske, A., Gekeler, V., *et al.* (2008). Histone deacetylases 1, 2 and 3 are highly expressed in prostate cancer and HDAC2 expression is associated with shorter PSA relapse time after radical prostatectomy, *Br. J. Cancer* 98: 604–610.

Wells, A., Kassis, J., Solava, J., *et al.* (2002). Growth factor-induced cell motility in tumor invasion, *Acta Oncol.* **41(2)**: 124–130.

Welsbie, D. S., Xu, J., Chen, Y., *et al.* (2009). Histone deacetylases are required for androgen receptor function in hormone-sensitive and castrate-resistant prostate cancer, *Cancer Res.* **69**: 958–986.

Westin, P., Bergh, A., and Damber, J. E. (1993). Castration rapidly results in a major reduction in epithelial cell numbers in the rat prostate, but not in the highly differentiated Dunning R3327 prostatic adenocarcinoma, *Prostate* **22(1)**: 65.

Westin, P., Stattin, P., Damber, J. E., *et al.* (1995). Castration therapy rapidly induces apoptosis in a minority and decreases cell proliferation in a majority of human prostatic tumors, *Am. J. Pathol.* **146(6)**: 1368.

Whang, Y. E., Wu, X., Suzuki, H., *et al.* (1998). Inactivation of the tumor suppressor PTEN/MMAC1 in advanced human prostate cancer through loss of expression, *Proc. Natl. Acad. Sci. USA* **95**: 5246–5250.

Whitmore, W. J. Jr (1984). Natural history and staging of prostate cancer, *Urol. Clin. North. Am.* **11**: 209–220.

Wickstrom, S. A., Alitalo, K., and Keski-Oja, J. (2002). Endostatin associates with integrin alpha5beta1 and caveolin-1, and activates Src via a tyrosyl phosphatase-dependent pathway in human endothelial cells, *Cancer Res.* **62(19)**: 5580–5589.

Wiechen, K., Sers, C., Agoulnik, A., *et al.* (2001). Down-regulation of caveolin-1, a candidate tumor suppressor gene, in sarcomas, *Am. J. Pathol.* **158(3)**: 833–839.

Wikstrom, P., Stattin, P., Franck-Lissbrant, I., *et al.* (1998). Transforming growth factor beta1 is associated with angiogenesis, metastasis, and poor clinical outcome in prostate cancer, *Prostate* **37**: 19–29.

Wikstrom, P., Westin, P., Stattin, P., *et al.* (1999). Early castration-induced upregulation of transforming growth factor beta1 and its receptors is associated with tumor cell apoptosis and a major decline in serum prostate-specific antigen in prostate cancer patients, *Prostate* **38(4)**: 268–277.

Williams, T. M., Hassan, G. S., Li, J., *et al.* (2005). Caveolin-1 promotes tumor progression in an autochthonous mouse model of prostate cancer: genetic ablation of Cav-1 delays advanced prostate tumor development in TRAMP mice, *J. Biol. Chem.* **280(26)**: 25134–25145.

Winter, R. N., Kramer, A., Borkowski, A., *et al.* (2001). Loss of caspase-1 and caspase-3 expression in human prostate cancer, *Cancer Res.* **61**: 1227–1232.

Winter, R. N., Rhee, J., and Kyprianou, N. (2004). Caspase-1 enhances the apoptotic sensitivity of human prostate cancer cells to ionizing irradiation, *Anticancer Research* **24**: 1377–1386.

Wolf, B. B., and Green, D. R. (1999). Suicidal tendencies: apoptotic cell death by caspase family proteinases, *J. Biol. Chem.* **274**: 20049–20052.

Wolk, A., Mantzoros, C. S., Andersson, S. O., *et al.* (1998). Insulin-like growth factor 1 and prostate cancer risk: a population-based, case-control study, *J. Natl. Cancer Inst.* **90(12)**: 911–915.

Wright, A. S., Douglas, R. C., Thomas, L. N., *et al.* (1999). Androgen-induced regrowth in the castrated rat ventral prostate: role of 5alpha-reductase, *Endocrinology* **140(10)**: 4509–4515.

Wu, L., Birle, D. C., and Tannock, I. F. (2005). Effects of the mammalian target of rapamycin inhibitor CCI-779 used alone or with chemotherapy on human prostate cancer cells and xenografts, *Cancer Res.* **65**: 2825–2831.

Wu, R., Zhai, Y., Fearon, E. R., and Cho, K. R. (2001). Diverse mechanisms of β-catenin deregulation in ovarian endometrioid adenocarcinomas, *Cancer Res.* **61**: 8247.

Wu, Y., Zhao, W., Zhao, J., *et al.* (2007). Identification of androgen response elements in the insulin-like growth factor I upstream promoter, *Endocrinology* **148(6)**: 2984–2993.

Wu, Z., Chang, P-C., Yang, J. C., *et al.* (2010). Autophagy blockade sensitizes prostate cancer cells toward SRC family kinase inhibitors, *Genes & Cancer* **1**: 40–49.

Wyllie, A. H., Kerr, J. F., and Currie, A. R. (1980). Cell death: the significance of apoptosis, *Int. Rev. Cytol.* **68**: 251.

Xie, S., Lin, H. K., Ni, J., *et al.* (2004). Regulation of interleukin-6-mediated PI3K activation and neuroendocrine differentiation by androgen signaling in prostate cancer LNCaP cells, *Prostate* **60**: 61–67.

Xue, C., Wyckoff, J., Liang, F., *et al.* (2006). Epidermal growth factor receptor overexpression results in increased tumor cell motility *in vivo* coordinately with enhanced intravasation and metastasis, *Cancer Res.* **66(1)**: 192–197.

Yamaguchi, H., and Wang, H. G. (2004). CHOP is involved in endoplasmic reticulum stress-induced apoptosis by enhancing DR5 expression in human carcinoma cells, *J. Biol. Chem.* **279(44)**: 45495–45502.

Yamamoto, K., Sato, T., Matsui, T., *et al.* (2007). Transcriptional induction of mammalian ER quality control proteins is mediated by single or combined action of ATF6 and XBP1, *Dev. Cell.* **13(3)**: 365–376.

Yang, E., Zha, J., Jockel, J., *et al.* (1995). BAD, a heterodimeric partner for Bcl-x$_L$ and Bcl-2, displaces bax and promotes cell death, *Cell* **80(2)**: 285–291.

Yang, G., Truong, L. D., Timme, T. L., *et al.* (1998). Elevated expression of caveolin is associated with prostate and breast cancer, *Clin. Cancer Res.* **4(8)**: 1873–1880.

Yang, J., Mani, S. A., Donaher, J. L., *et al.* (2004). Twist, a master regulator of morphogenesis, plays an essential role in tumor metastasis, *Cell* **117(7)**: 927–939.

Yegnasubramanian, S., Haffner, M. C., Zhang, Y., *et al.* (2008). DNA hypomethylation arises later in prostate cancer progression than CpG island hypermethylation and contributes to metastatic tumor heterogeneity, *Cancer Res.* **68**: 8954–8967.

Yegnasubramanian, S., Kowalski, J., Gonzalgo, M. L., *et al.* (2004). Hypermethylation of CpG islands in primary and metastatic human prostate cancer, *Cancer Res.* **64**: 1975–1986.

Yeh, S., Lin, H. K., Kang, H. Y., *et al.* (1999). From HER2/Neu signal cascade to androgen receptor and its coactivators: a novel pathway by induction of androgen target genes through MAP kinase in prostate cancer cells, *Proc. Natl. Acad. Sci. USA* **96(10)**: 5458–5463.

Yin, C., Knudson, C. M., Korsmeyer, S. J., *et al.* (1997). BAX suppresses tumorigenesis and stimulates apoptsis *in vivo*, *Nature* **385**: 637–640.

Yingling, J. M., Blanchard, K. L., and Sawyer, J.S. (2004). Development of TGF-β signalling inhibitors for cancer therapy, *Nat. Rev. Drug Discov.* **3**: 1011–1022.

Yoneda, T., and Hiraga, T. (2005). Crosstalk between cancer cells and bone microenvironment in bone metastasis, *Biochem. Biophys. Res. Commun.* **328(3)**: 679–687.

Yoneda, T., Imaizumi, K., Oono, K., *et al.* (2001). Activation of caspase-12, an endoplasmic reticulum (ER) resident caspase, through tumor necrosis factor receptor-associated factor 2-dependent mechanism in response to the ER stress, *J. Biol. Chem.* **276**: 13935–13340.

Yoon, G., Kim, J. Y., Choi, Y. K., *et al.* (2006). Direct activation of TGF-beta1 transcription by androgen and androgen receptor complex in Huh7 human hepatoma cells and its tumor in nude mice, *J. Cell Biochem.* **97(2)**: 393–411.

Yoshida, B. A., Sokoloff, M. M., Welch, D. R., *et al.* (2000). Metastasis-suppressor genes: a review and perspective on an emerging field, *J. Natl. Cancer Inst.* **92(21)**: 1717–1730.

Yoshida, H., Kong, Y., Yoshida, R., *et al.* (1998). Apaf1 is required for mitochondrial pathways of apoptosis and brain development, *Cell* **94(6)**: 739–750.

Yoshida, H., Matsui, T., Yamamoto, A., *et al.* (2001). XBP1 mRNA is induced by ATF6 and spliced by IRE1 in response to ER stress to produce a highly active transcription factor, *Cell* **107(7)**: 881–891.

Yoshizawa, A., and Ogikubo, S. (2006). IGF binding protein-5 synthesis is regulated by testosterone through transcriptional mechanisms in androgen responsive cells, *Endocr. J.* **53(6)**: 811–818.

Yoshizawa, K., Willett, W. C., Morris, S. J., *et al.* (1998). Study of prediagnostic selenium level in toenails and the risk of advanced prostate cancer, *J. Natl. Cancer Inst.* **90**: 1219–1224.

Yu, F., Finley, R. L. Jr, Raz, A., *et al.* (2002). Galectin-3 translocates to the perinuclear membranes and inhibits cytochrome c release from the mitochondria. A role for synexin in galectin-3 translocation, *J. Biol. Chem.* **277(18)**: 15819–15827.

Yu, J., Zhang, L., Hwang, P. M., *et al.* (2001). PUMA induces the rapid apoptosis of colorectal cancer cells, *Mol. Cell* **7**: 673–682.

Yuan, J., Shaham, S., Ledoux, S., *et al.* (1993). The *C. elegans* cell death gene *ced-3* encodes a protein similar to mammalian interleukin-1 beta-converting enzyme, *Cell* **75(4)**: 641.

Yuan, X., and Balk, S. P. (2009). Mechanisms mediating androgen receptor reactivation after castration, *Urol. Oncol.* **27(1)**: 36–41.

Yuan, X. J., and Whang, Y. E. (2002). PTEN sensitizes prostate cancer cells to death receptor-mediated and drug-induced apoptosis through a FADD-dependent pathway, *Oncogene* **21(2)**: 319–327.

Yue, Z., Jin, S., Yang, C., *et al.* (2003). Beclin 1, an autophagy gene essential for early embryonic development, is a haploinsufficient tumor suppressor, *Proc. Natl. Acad. Sci. USA* **100**: 15077–15082.

Zabrenetzky, V., Harris, C. C., Steeg, P. S., *et al.* (1994). Expression of the extracellular matrix molecule thrombospondin inversely correlates with malignant progression in melanoma, lung and breast carcinoma cell lines, *Int. J. Cancer* **59(2)**: 191–195.

Zagars, G. K., Pollack, A., Kavadi, V. S., *et al.* (1995). PSA and radiation therapy for clinically localized prostate cancer, *Int. J. Rad. Oncol. Biol. Phys.* **32**: 293.

Zalckvar, E., Berissi, H., Eisenstein, M., *et al.* (2009a). Phosphorylation of Beclin 1 by DAP-kinase promotes autophagy by weakening its interactions with Bcl-2 and BCL-XL, *Autophagy* **5**: 720–722.

Zalckvar, E., Berissi, H., Mizrachy, L., *et al.* (2009b). DAP-kinase-mediated phospho-rylation on the BH3 domain of beclin 1 promotes dissociation of beclin 1 from BCL-X$_L$ and induction of autophagy, *EMBO Rep.* **10**: 285–292.

Zavadil, J., and Bottinger, E. P. (2005). TGF-beta and epithelial-to-mesenchymal transi-tions, *Oncogene* **24**(37): 5764–5774.

Zeng, L., Rowland, R. G., Lele, S. M., *et al.* (2004). Apoptosis incidence and protein expression of p53, TGF-beta receptor II, p27Kip1, and Smad4 in benign, premali-gnant, and malignant human prostate, *Hum. Pathol.* **35**(3): 290–297.

Zerbini, L. F., Wang, Y., Cho, J. Y., *et al.* (2003). Constitutive activation of nuclear factor kappaB p50/p65 and Fra-1 and JunD is essential for deregulated interleukin 6 expression in prostate cancer, *Cancer Res.* **63**(9): 2206–2215.

Zha, S., Gage, W. R., Sauvageot, J., *et al.* (2001). Cyclooxygenase-2 is up-regulated in proliferative inflammatory atrophy of the prostate, but not in prostate carcinoma, *Cancer Res.* **61**: 8617–8623.

Zhan, M., Zhao, H., and Han, Z. C. (2004). Signaling mechanisms of anoikis, *Histol. Histopathol.* **19**(3): 973–983.

Zhang, M., Magit, D., and Sager, R. (1997). Expression of maspin in prostate cells is regulated by a positive ets element and a negative hormonal responsive element site recognized by androgen receptor, *Proc. Natl. Acad. Sci. USA* **94**(11): 5673–5678.

Zhang, M., Volpert, O., Shi, Y. H., *et al.* (2000). Maspin is an angiogenesis inhibitor, *Nat. Med.* **6**(2): 196–199.

Zhang, X., Jin, T. G., Yang, H., *et al.* (2004a). Persistent c-FLIP(L) expression is necessary and sufficient to maintain resistance to tumor necrosis factor-related apoptosis-inducing ligand-mediated apoptosis in prostate cancer, *Cancer Res.* **64**(19): 7086–7091.

Zhang, Y., Lu, H., Dazin, P., *et al.* (2004b). Squamous cell carcinoma cell aggregates escape suspension-induced, p53-mediated anoikis: fibronectin and integrin alphav mediate survival pathways in cells, *J. Biol. Chem.* **279**: 48342–48349.

Zhang, Y., Nan, B., Yu, J., *et al.* (2002a). From castration-induced apoptosis of prostatic epithelium to the use of apoptotic genes in the treatment of prostate cancer, *Ann. N. Y. Acad. Sci.* **963**: 191.

Zhang, Y., Ni, J., Messing, E. M., *et al.* (2002b). Vitamin E succinate inhibits the function of androgen receptor and the expression of prostate-specific antigen in prostate cancer cells, *Proc. Natl. Acad. Sci. USA* **99**: 7408–7413.

Zhao, Y., Burikhanov, R., Qiu, S., *et al.* (2007). Cancer resistance in transgenic mice expressing the SAC module of Par-4, *Cancer Res.* **67**: 9276–9285.

Zhao, Y., and Rangnekar, V. M. (2008). Apoptosis and tumor resistance conferred by par-4, *Cancer Biol. Ther.* **7**(12): 1867–1874.

Zhau, H. E., Odero-Marah, V., Lue, H-W., *et al.* (2008). Epithelial to mesenchymal transition (EMT) in human prostate cancer: lessons learned from ARCaP model, *Clin. Exp. Metastas.* **25**(6): 601–610.

Zhou, B. P., Deng, J., Xia, W., *et al.* (2004). Dual regulation of Snail by GSK-3beta-mediated phosphorylation in control of epithelial-mesenchymal transition, *Nat. Cell Biol.* **6**(10): 931–940.

Zhou, J., Zhong, D-W., Wang, Q-W., *et al.* (2010). Paclitaxel ameliorates fibrosis in hepatic stellate cells via inhibition of TGF-β/Smad activity, *World J. Gastroenterol.* **16**(26): 3330–3334.

Zhu, B., Fukada, K., Zhu, H., *et al.* (2006). Prohibitin and cofilin are intracellular effectors of transforming growth factor beta signaling in human prostate cancer cells, *Cancer Res.* **66**(17): 8640–8647.

Zhu, B., and Kyprianou, N. (2005). Transforming growth factor beta and prostate cancer, *Cancer Treat. Res.* **126**: 157–173.

Zhu, B., Zhai, J., Zhu, H. *et al.* (2010a). Prohibitin regulates TGF-β induced apoptosis as a downstream effector of Smad-dependent and -independent signaling, *Prostate* **70**: 17–26.

Zhu, M., and Kyprianou, N. (2010). Role of androgens and the androgen receptor in epithelial-mesenchymal transition and invasion of prostate cancer cells, *FASEB J.* **24**: 769–777.

Zhu, M. L., Horbinski, C. M., Garzotto, M., *et al.* (2010b). Tubulin-targeting chemotherapy impairs androgen receptor activity in prostate cancer, *Cancer Res.* **70**(20): 7992–8002.

Zhu, M. L., and Kyprianou, N. (2008). Androgen receptor and growth factor signaling partnerships in prostate cancer cells, *Endocr. Relat. Cancer* **15**: 841–849.

Zhu, M. L., Partin, J. V., Strup, S. E., *et al.* (2008). TGF-β signaling and androgen receptor status determine apoptotic cross-talk in human prostate cancer cells, *Prostate* **68**(3): 287–295.

Zhu, Z., Sanchez-Sweatman, O., Huang, X., *et al.* (2001). Anoikis and metastatic potential of cloudman S91 melanoma cells, *Cancer Res.* **61**(4): 1707–1716.

Zhuang, L., Kim, J., Adam, R. M., *et al.* (2005). Cholesterol targeting alters lipid raft composition and cell survival in prostate cancer cells and xenografts, *J. Clin. Invest.* **115**(4): 959–968.

Zietman, A. L., Coen, J. J., Shipley, W. U., *et al.* (1994). Radical radiation therapy in the management of prostatic adenocarcinoma: the initial prostate specific antigen value as a predictor of treatment outcome, *J. Urol.* **151**: 640–645.

Zong, W. X., Li, C., Hatzivassiliou, G., *et al.* (2003). BAX and BAK can localize to the endoplasmic reticulum to initiate apoptosis, *J. Cell Biol.* **162**(1): 59–69.

Zong, W. X., Lindsten, T., Ross, A. J., *et al.* (2001). BH3-only proteins that bind pro-survival BCL-2 family members fail to induce apoptosis in the absence of Bax and Bak, *Genes Dev.* **15**: 1481–1486.

Zou, Z., Anisowicz, A., Hendrix, M. J., *et al.* (1994). Maspin, a serpin with tumor-suppressing activity in human mammary epithelial cells, *Science* **263**(5146): 526–529.

Zu, K., and Ip, C. (2003). Synergy between selenium and vitamin E in apoptosis induction is associated with activation of distinctive initiator caspases in human prostate cancer cells, *Cancer Res.* **63**: 6988–6995.

Zuniga, A., Quillet, R., Perrin-Schmitt, F., *et al.* (2002). Mouse Twist is required for fibroblast growth factor-mediated epithelial-mesenchymal signalling and cell survival during limb morphogenesis, *Mech. Dev.* **114**(1–2): 51–59.

# Abbreviations

AIF, Apoptosis-inducing factor

AKT/PKB, protein kinase B

AMPK, 5' adenosine monophosphate-activated protein kinase

Ang, angiopoietin

APAF1, Apoptotic protease-activating factor 1

AR, Androgen receptor

ARE, Androgen response element

BAK, BCL-2 antagonist or killer

BAX, BCL-2 associated X protein

BCL-2, B-cell lymphoma 2

bFGF, Basic fibroblast growth factor

BH, BCL-2 homology regions

BIM, BCL-2 interacting mediator of cell death

CEP, circulating endothelial precursor cells

COX-2, cyclooxygenase-2

CRPC, castration-resistant prostate cancer

$5\alpha$-DHT, 5 alpha-dihydrotestosterone

DBD, DNA-binding domain

DED, Death effector domain

DISC, Death-inducing signaling complex

ECM, Extracellular matrix

EGF, Epidermal growth factor

EGFR, Epidermal growth factor receptor

EMT, Epithelial–mesenchymal-transition

ER, Estrogen receptor
ERK, extracellular signal-regulated kinase
ESR, endoplasmic reticulum stress response
ET-1, Endothelin-1
FADD, Fas-associated death domain protein
FAK, Focal adhesion kinase
FASL, Fas ligand
FDA, Food and Drug Administrations
FLIP, FLICE-inhibitory protein
GM-CSF, granular-macrophage colony-stimulating factor
GSK3, glycogen synthase kinase-3
GSTP1, glutathionine S-transferase
HATs, histone acetyltransferases
HDACs inhibitors, histone deacetylase inhibitors
HGF, hepatocyte growth factor
HIF-1$\alpha$, hypoxia-inducible factor alpha
HUVECs, human vascular endothelial cells
IAP, inhibitor of apoptosis proteins
IGF-1, Insulin growth factor-1
IGF1R, IGF receptor
IGFBP, Insulin growth factor receptor binding protein
IKK, I$\kappa$B Kinase complex
IL-3, Interleukin 3
IL-6, Interleukin 6
ILK, Integrin-linked kinase
IMS, Intermembrane space proteins
JNK, JUN N-terminal kinase
LBD, Ligand binding domain
LOX, Lysyl oxidase
MEK, mitogen and extracellular kinase
MMPs, matrix metalloproteinases
MOMP, Mitochondrial outer membrane permeabilization
NF-$\kappa$B, nuclear factor kappa B
NGF, Nerve growth factor
NSAIDs, Non-steroidal anti-inflammatory drugs
NTD, Amino terminal domain

PI3K, phosphoinositide 3-kinase

MAPK, mitogen-activated protein kinase

PAR-4, Prostate apoptosis response-4

PARP, Poly ADP-ribose polymerase

PCTP, Prostate Cancer Prevention Trial

PDGF, platelet-derived growth factor

PKC, protein kinase C

PSA, Prostate-specific antigen

REDUCE, Reduction by Dutasteride of Prostate Cancer Events

ROS, Reactive oxygen species

SERCA, sarcoplasmic/endplasmic reticulum calcium ATPase

SMAC/DIABLO, second mitochondria-derived activator of caspases

TAMs, Tumor associated macrophages

TGF-β, transforming growth factor-beta

TGFβRI, transforming growth factor-beta receptor I

TGFβRII, transforming growth factor-beta receptor II

TNF-α, Tumor Necrosis Factor alpha

TNFR, Tumor necrosis factor receptor

TORC1, target of rapamycin (TORC1) inhibitor

TRADD, TNF-R1-associated death domain containing protein

TRAIL, TNF-related apoptosis-inducing ligand

TRAMP, Transgenic adenocarcinoma mouse prostate

TSP-1, Thrombospondin-1 TSP-1

VEGF, vascular endothelial growth factor

VEGFR2, vascular endothelial growth factor receptor 2

XIAP, X-linked inhibitor of apoptosis protein

# Index

# Acknowledgements

Dr John T. Isaacs (Johns Hopkins University), for his generous mentoring in the mysteries of prostate cancer and cell death, at the very beginning of my research career. Dr Stephen C. Jacobs (University of Maryland), Dr Randall G. Rowland (University of Kentucky), and the legendary Dr James Glenn, three of the greatest urologists of our time, for their leadership, support, and friendship. Dr Craig Horbinski, Dr Shinichi Sakamoto, Dr MengLei Zhu, Sarah Martin, and Lorie Howard, members of my research team, for their creative skills and help in the preparation of the figures and organization of the bibliography for this monograph. The constant support and friendship of my colleagues (at the University of Kentucky and Institutions around the globe), my friends and family have been truly invaluable throughout the journey. My father, George Kyprianou, died before seeing this book finished, but his constant encouragement and faith in me remain the most divine and inspirational gifts. He was the greatest human being I have ever known, enriching my life every step of the way.

www.ingramcontent.com/pod-product-compliance
Lightning Source LLC
Chambersburg PA
CBHW050559190326
41458CB00007B/2099